High Performance Tensegrity-Inspired Metamaterials and Structures

Following current trends toward development of novel materials and structures, this volume explores the concept of high-performance metamaterials and metastructures with extremal mechanical properties, inspired by tensegrity systems.

The idea of extremal materials is applied here to cellular tensegrity lattices of various scales. Tensegrity systems have numerous advantages: they are lightweight, have a high stiffness-to-mass ratio, are prone to structural control, can be applied in smart and adaptive systems, and exhibit unusual mechanical properties. This study is focused on tensegrity lattices, whose inner architecture resembles that of cellular metamaterials, but which are aimed at civil engineering applications in non-material scales. It proposes a methodology for investigation of extremal mechanical properties of such systems, based on discrete and continuum approaches, including the discussion on scale effects. It proves that, similarly to tensegrity-based metamaterials, tensegrity metastructures are able to exhibit extremal mechanical behaviour.

This book is directed to researchers and scientists working on metamaterials and tensegrity systems, developing energy-absorption solutions for building and transport industry. The findings described in this monograph can also be useful in other fields of applied sciences, such as civil engineering, robotics, and material science.

High Performance Tensegrity-Inspired Metamaterials and Structures

Anna Al Sabouni-Zawadzka

CRC Press
Taylor & Francis Group
Boca Raton London New York

CRC Press is an imprint of the
Taylor & Francis Group, an **informa** business

First edition published 2023
by CRC Press
6000 Broken Sound Parkway NW, Suite 300, Boca Raton, FL 33487-2742

and by CRC Press
4 Park Square, Milton Park, Abingdon, Oxon, OX14 4RN

CRC Press is an imprint of Taylor & Francis Group, LLC

ISBN: 978-1-032-38041-4 (hbk)
ISBN: 978-1-032-38042-1 (pbk)
ISBN: 978-1-003-34320-2 (ebk)

DOI: 10.1201/9781003343202

Typeset in LM Roman 10
by KnowledgeWorks Global Ltd.

*Publisher's note:*This book has been prepared from camera-ready copy provided by the authors.

Dedication

_____

To Adam and Franek

Contents

Preface

The past few decades have brought a rapid development of engineering solutions, such as novel metamaterials with extraordinary characteristics, smart materials and structures, adaptive systems with adjustable properties or extremal materials. Scientists have been searching for systems with enhanced behaviour and unique properties, which could be advantageous in emerging new applications, among others in civil engineering.

In this book, the concept of extremal materials is applied to cellular tensegrity lattices of various scales. Tensegrity systems have numerous advantages: they are lightweight, have a high stiffness-to-mass ratio, are prone to structural control, can be applied in smart and adaptive systems, offer wide possibilities of geometry shaping, and exhibit unusual mechanical properties.

This study is focussed on tensegrity lattices, whose inner architecture resembles that of cellular metamaterials, but which are aimed at civil engineering applications in non-material scales. I have proposed a methodology for investigation of extremal mechanical properties of such systems, including the discussion on scale effects. I have proved that similarly to tensegrity-based metamaterials, tensegrity metastructures may exhibit extremal mechanical behaviour.

The first chapter, Introduction, introduces the reader to the concept of high-performance tensegrity-based metamaterials and structures, explaining the ideas of tensegrity systems, mechanical metamaterials, and extremal materials. In the second chapter, Extremal Materials and Structures, many examples of mechanical metamaterials and metastructures with extremal mechanical properties are presented. The third chapter, Tensegrity—Smart Structures with Unusual Mechanical Properties, is focussed on tensegrity systems with their inherent smart properties. The fourth chapter, Analysis of Tensegrity Systems, introduces two approaches to the analysis of tensegrity-based mechanical metamaterials and metastructures, namely discrete and continuum models. The most important part of this book is its fifth chapter, Extremal Mechanical Properties of Tensegrity Systems, where very thorough analyses of extremal mechanical behaviour of a series of 2D and 3D tensegrity modules and lattices are presented. This chapter is important, as it introduces a novel methodology for evaluating extremal mechanical properties of tensegrity metastructures, based on a consecutive use of continuum and discrete approaches. The last chapter, Technology and Applications, discusses technological aspects and potential applications of tensegrity lattices with extremal mechanical behaviour in the field of civil engineering.

This book is directed to researchers and scientists working on metamaterials and tensegrity systems, developing energy-absorption solutions for building and transport industry. However, the findings described in this monograph can also be useful in other fields of applied sciences, e.g. civil engineering, robotics, material science. I have proposed solutions that have many potential application areas—the most important are energy absorption systems, but they can also be applied, for example, as 3D fillings of conventional structural elements with the aim of enhancing their mechanical behaviour.

Finally, I would like to express my gratitude to Professor Wojciech Gilewski for his continuous support, his time, and involvement. He introduced me to the research on tensegrity systems and has supported me every step of the way.

Author Biography

Anna Al Sabouni-Zawadzka is Assistant Professor at the Warsaw University of Technology, Faculty of Civil Engineering, Poland. She specialises in the research on tensegrity systems.

1 Introduction

The last years have brought development of numerous novel engineering solutions, such as metamaterials with extraordinary properties, smart materials and structures, adaptive systems with adjustable behaviour or extremal materials. Researchers have studied systems with enhanced behaviour and unique properties, which could be advantageous in emerging new applications, among others in civil engineering.

Among structural systems with a potential to be used in such innovative engineering solutions, tensegrity structures are the ones that deserve special attention. The concept of tensegrity was introduced by Buckminster Fuller [52] and Kenneth Snelson [126], who used a combination of two words: *tension* and *integrity* to describe a structural principle, where a system of isolated compressed members remains inside a continuous net of elements in tension. One of the most characteristic features of tensegrity is the occurrence of infinitesimal mechanisms, which are balanced with self-stress states. While finite mechanisms describe an arbitrary geometrical instability of the structure, infinitesimal mechanisms are related to the local geometrical instability in the range of infinitesimal displacements.

Tensegrity systems have numerous advantages: they are lightweight, have a high stiffness-to-mass ratio, are prone to structural control, can be applied in smart and adaptive systems, offer wide possibilities of geometry shaping, and exhibit unusual mechanical properties. Moreover, tensegrities have all the inherent attributes of smart structures, which include: self-control, self-diagnosis, self-repair, and active control. Adjustment of prestressing forces in cables and struts makes it possible to change mechanical properties of these unique structures and influence their behaviour.

Apart from macro-scale applications in civil and mechanical engineering, tensegrity systems can be used to construct cellular mechanical metamaterials and lattices in various scales. Metamaterials are artificial composite systems with extraordinary properties, which result mainly from the morphology of the structure and, to a smaller degree, from chemical or phase composition. Within metamaterials with a cellular architecture, structural lattices are of particular interest in this study. Depending on the applied scale, lattices can be regarded as materials or structures.

Cellular metamaterials can be designed in such a way that they exhibit a variety of unusual properties, such as for example: a negative Poisson's ratio, a vanishing shear modulus, negative compressibility or negative stiffness, as well as extremal mechanical properties. The concept of extremal materials was introduced in 1995 by Graeme Milton and Andrej Cherkaev [94]. They should be understood as systems with extremely large stiffness in some modes of deformation and extremely compliant behaviour in others.

DOI: 10.1201/9781003343202-1   1

The study of extremal mechanical properties of structural systems is based on the analysis of their elasticity tensor. As it is known, it must be positive definite, and in the theory of elasticity it shows certain types of symmetries. This tensor can be diagonalised by orthogonal transformation. If we present the components of the elastic tensor in the Voigt notation as a square matrix $\mathbf{E}$ of dimensions 6×6, its diagonal representation is a set of eigenvalues $\lambda_i > 0$ ($i = 1, 2, \ldots, 6$), and orthogonal eigenvectors $\mathbf{w}_i$ describe modes of deformation. One can classify systems as: nullmode, unimode, bimode, trimode, quadramode, pentamode, or hexamode, depending on the number of eigenvalues λ_i that approach zero.

In this book, the concept of extremal materials, which was described by Milton and Cherkayev [94] within the material scale, is applied to tensegrity metamaterials as well as bigger structural systems, such as cellular lattices, whose inner architecture resembles that of cellular metamaterials, but which are aimed at civil engineering applications in a non-material scale (they are further referred to as metastructures). It is shown that tensegrities, thanks to their unique mechanical properties, can be used to construct high-performance systems with extremal behaviour, not only in a material scale, but also in a structural one.

The book consists of six chapters:

Chapter 1 is the introduction to the study on high-performance tensegrity-based metamaterials and structures. It introduces tensegrity systems, mechanical metamaterials, and extremal materials.

Chapter 2 discusses mechanical metamaterials and metastructures with extremal mechanical properties, focusing on definitions, a literature review, and description of unusual mechanical properties of the systems. It gives many examples of high-performance structural systems, including adaptive origami- and tensegrity-inspired metamaterials and structures.

Chapter 3 focuses on tensegrity systems and their smart properties. It introduces the concept of smart systems, and presents examples of numerical analyses which prove that tensegrities exhibit a series of inherent properties, and therefore, can be regarded as smart structural systems.

Chapter 4 presents two approaches to the analysis of tensegrity metamaterials and metastructures, namely discrete and continuum models. Within each approach, a step-by-step procedure for the calculation of tensegrity systems is given. It also discusses scale effects in the continuum approach.

Chapter 5 is the most important part of this book, as it contains detailed analyses of extremal behaviour of a series of 2D and 3D tensegrity modules and lattices. It presents a methodology for identifying extremal mechanical properties of metamaterials within the continuum model, and proposes a methodology for evaluating extremal behaviour of tensegrity metastructures, based on a consecutive application of continuum and discrete approaches.

Chapter 6 discusses technological aspects and potential applications of the proposed tensegrity lattices in civil engineering, both in a material and non-material scale.

2 Extremal Materials and Structures

2.1 METAMATERIALS

Metamaterials are usually defined as human-designed and human-made, not observed in nature, composite structures with unusual or non-typical properties [36,41,124]. Features of metamaterials are determined mainly by morphology of the structure and, to a lesser degree, by chemical or phase composition. Metamaterials can be applied in various fields of engineering, including electromagnetism [121, 128], solar photovoltaic systems [142], energy absorption systems [24,34], etc. Further, the author will focus on one special class of metamaterials, namely, cellular mechanical metamaterials [21,71,81,82,151]. They are artificial structures with unique properties, which result from their inner architecture. Such metamaterials exhibit extraordinary mechanical properties, for example negative Poisson's ratio, non-typical modulus of extension and volumetric changes, ultra-light and ultra-stiff behaviour, unusual dynamic properties.

Mechanical metamaterials and their engineering applications have been under considerable and important scientific research in recent years [15,130, 143, 145]. An inspiration for first cellular metamaterials came from nature [130]. Natural cellular materials have evolved over millions of years to develop optimal architecture and unique properties. A good example is a complex structure of a bone core, which consists of intricately shaped ligaments with variable density to achieve a very high structural efficiency [130]. Artificial cellular materials are not a new concept, however, thanks to the recent developments in additive manufacturing techniques (such as a 3D printing), the fabrication of metamaterials with complex inner structures has become possible within a wide range of scales: from nano- to macro-structures.

Cellular metamaterials consist of an interconnected network of unit cells, which are usually created from 1D (e.g. struts) and 2D (e.g. plates) structural elements. Their main function is to enhance the mechanical behaviour and to adjust it to the desired application. Mechanical properties of such materials depend on numerous factors, such as properties of the parent material of the cells, topology of the system, relative density, applied scale, structural properties of the cells, etc. A proper design of the mechanical metamaterial can lead to novel extraordinary properties, which are observed neither in nature nor in traditional artificial materials. For instance, it is possible to obtain materials with negative Poisson's ratio, adjustable stiffness, negative compressibility or extremal mechanical properties, which are discussed in detail in this study.

DOI: 10.1201/9781003343202-2

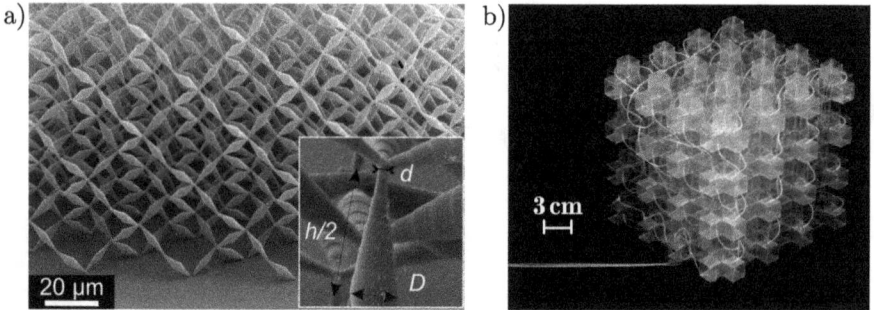

Figure 2.1 Examples of mechanical metamaterials: a) polymer pentamode metamaterial (reproduced from M. Kadic, T. Bückmann, N. Stenger, M. Thiel, and M. Wegener. On the practicability of pentamode mechanical metamaterials. Applied Physics Letters, 100(19):191901, 2012, with the permission of AIP Publishing); b) origami-inspired metamaterial (reproduced from J. Overvelde, T. de Jong, Y. Shevchenko, S. Becerra, G. Whitesides, J. Weaver, C. Hoberman, and K. Bertoldi. A three-dimensional actuated origami-inspired transformable metamaterial with multiple degrees of freedom. Nature Communications, 7(1), 2016, under the Creative Commons Attribution 4.0 International License, added scale bar.)

Cellular metamaterials can be divided into two categories: closed-cell and open-cell materials [130], with two main types of deformation behaviour: bending-dominated and stretching-dominated. While the open-cell systems have a microstructure consisting of a network of interconnected struts that fill a 3D space, the closed-cell materials contain plate-like structures with a defined thickness and length. Within the open-cell systems, a sub-category of lattice metamaterials can be distinguished. These artificial cellular materials consist of a large number of rods or slender beams and have a regular pattern created by unit cells periodically repeated in space.

Figure 2.1 presents examples of two 3D cellular metamaterials. Figure 2.1a shows a pentamode mechanical metamaterial with ultra-large bulk to shear modulus ratio manufactured using the Two-Photon Lithography (TPL) [71]. In Figure 2.1b a transformable origami-inspired mechanical metamaterial actuated by pressurising its unit cells is shown [104]. It can be noticed that the presented examples cover various length scales, with structural elements ranging from a few micrometers to a few millimeters.

Depending on the applied scale, lattices can be regarded as metamaterials or metastructures. The lattice should be treated as a material when the size of a unit cell is very small compared to the size of the whole system. The main focus of this study are lattices, which have the inner architecture of metamaterials, but are to be applied as structural systems in civil engineering scale. Further in this study, they will be referred to as lattices or metastructures. Such systems, however, cannot be considered nor analysed in isolation from metamaterial structures, as numerous similarities occur within the considered scale ranges, what is discussed in Chapters 4 and 5.

2.2   EXTREMAL MATERIALS

The idea of extremal materials was introduced by Milton and Cherkayev in 1995 [94]. The term *extremal* implies that the material is extremely stiff in certain modes of deformation and extremely compliant in others. Some of extremal materials exhibit unusual mechanical properties, such as for example a negative Poisson's ratio. Extremal material properties may be examined by analysing the elasticity tensor, which can be diagonalised using the orthogonal transformation. After diagonalisation, a set of eigenvalues is obtained, with orthogonal eigenvectors describing the deformation modes of the analysed material. A very small eigenvalue (i.e. approaching zero), indicates that the material behaviour is very compliant when subjected to deformation described by the corresponding eigenvector. Extremal materials can be classified depending on the number of eigenvalues of the elasticity tensor that are close to zero. The following materials can be distinguished: nullmode, unimode, bimode, trimode, quadramode, pentamode, or hexamode [71,94]. The nullmode material is an extremely stiff isotropic material, while the hexamode material is an extremely compliant isotropic material. The unimode material has one, bimode—two, trimode—three, quadramode—four, and pentamode—five compliant modes of deformation, further referred to as soft deformation modes.

What is very important from the point of view of mechanical metamaterials, Milton and Cherkayev [94] proved that extremal materials can be obtained by combining two specific parent materials—a very compliant and a very stiff one. This idea is illustrated in Figure 2.2. It can be easily noticed that a simple laminate consisting of two extremal parent materials is compliant under two types of loading and, at the same time, rigid under another one. Such extremal materials could afterwards be used to produce materials with any desired positive definite elasticity tensors. This finding implies that the extremal mechanical metamaterials may be fabricated by using additive

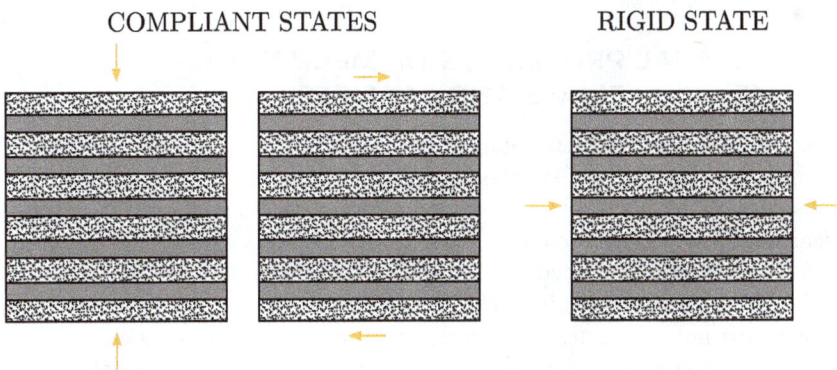

Figure 2.2 Example of a bimode material consisting of two extremal parent materials: a very compliant one (dotted region) and a very stiff one (grey solid region).

manufacturing techniques, namely by combining the void space (i.e. an extremely compliant material) with a very stiff 3D-printed material [145].

Following the study of Milton and Cherkayev [94], scientists have started to search for materials exhibiting extremal mechanical behaviour [26,39,71,92, 93,140]. One of the most thoroughly investigated metamaterials with extraordinary properties is a pentamode material [70,71,150], which was presented in Figure 2.1a. It is an extremely compliant system, whose elasticity tensor has only one non-zero eigenvalue. Therefore, isotropic metamaterials with a pentamode architecture can only withstand hydrostatic stress, which—together with their inherent water-like wave properties—makes them perfect for underwater acoustic applications [26].

Although pentamode materials have been widely studied in recent years, several other types of extremal metamaterials have also been found. Milton in [93] proposed 2D and 3D auxetic materials which can dilate to arbitrarily large strain with zero bulk modulus. In [92], on the other hand, he presented a family of non-linear bimode metamaterials constructed from rigid bars and pivots. Dudek et al. [39] analysed unimode metamaterials which are made from rotating rigid triangles, and exhibit negative linear compressibility and negative thermal expansion. Wei et al. [140] investigated wave properties of various quadramode metamaterials. Cai et al. [26] proposed a methodology for designing 2D extremal metamaterials using topology optimisation techniques, which can be extended to 3D tasks.

This study is an attempt to apply the concept of extremal materials, which was first introduced by Milton and Cherkayev [94], and afterwards investigated by other scientists [26, 39, 71, 140], to tensegrity-based systems. What is very important, the idea of extremal materials will be used here not only within a material scale, that is in tensegrity-based mechanical metamaterials, but will also be applied to bigger structural systems, such as metamaterial-inspired cellular lattices with extremal mechanical properties, aimed at civil engineering applications.

2.3   UNUSUAL PROPERTIES OF MECHANICAL METAMATERIALS AND LATTICES

A recent progress in the development of mechanical materials has lead to many interesting achievements, such as unusual mechanical properties or new functionalities of the materials. These include negative Poisson's ratio, vanishing shear modulus, negative compressibility, or negative stiffness [130,145].

Poisson's ratio is defined as a ratio of the strain in the direction transverse to the loading and the strain in the loading direction. For most traditional materials with homogeneous structure, the values of Poisson's ratio range from 0.0 to 0.5. In the case of mechanical metamaterials, however, local deformations around nodes may lead to a negative value of the Poisson's ratio [61]. Materials and structures that have a negative Poisson's ratio are called auxetics [15,130]. The word *auxetic* originates from the Greek term *auxeis*, which means *to increase and grow* [25]. Evans [43] was the first to use this term to describe the materials that expand in the direction perpendicular to the tensile loading. The global behaviour of auxetics differs from the local response of their constituents—they expand under tension and contract under compression.

One of the commonly known systems with negative Poisson's ratio are reentrant structures [144]. This unique feature was also discovered in origami-inspired metamaterials [22, 45], as well as tensegrity-based lattices [8]. The unusual behaviour of auxetic metamaterials results mainly from their inner architecture, namely folding and unfolding of their non-rigid elements. It influences other mechanical properties, which are favourable from the point of view of their potential applications, such as adjustable stiffness, high energy absorption, high indentation, and shear resistance [130]. In some of the mentioned auxetic metamaterials and lattices, it is even possible to adjust the Poisson's ratio, from negative to positive values, by modifying the geometry of their unit cells [8, 22].

Systems with negative Poisson's ratio exhibit a very good resistance against damage, as they are able to distribute stresses over a larger part of the material [25]. Thanks to their unique properties, they can be applied as shape memory materials, acoustic dampers, membrane filters, fasteners, etc. [25].

Another interesting property that can be obtained in mechanical metamaterials is a very large bulk modulus, which was observed in extremal pentamode metamaterials [130, 145]. These specific lattices, proposed by Milton and Cherkayev [94], have five compliant modes of deformation and thus, exhibit a large bulk modulus when compared to their shear modulus, which is close to zero. In practice, it means that the volumetric changes of pentamode materials are minimal and the Poisson's ratio is close to 0.5. In addition to large bulk modulus, there is an interesting phenomenon observed in pentamode lattices. Any stress transferred through the structure is transferred through the apex of the cones [70]. It means that the mass density and stiffness are not coupled—changes in the cone diameter do not affect the stiffness of the whole lattice. Example of a pentamode lattice, manufactured using the selective laser-melting technique at the Additive Manufacturing Laboratory of TU Delft, is presented in [145].

Negative compressibility is another unusual property that was discovered in mechanical metamaterials [15, 145]. Compressibility is related to the relative volume change of a material under the applied pressure and thus, it is typically a positive value [130]. Negative compressibility means that a system expands under the hydrostatic pressure and is only possible in specific materials and structures, such as some crystals, porous systems, composites, or certain lattices [16, 63, 64, 91].

Nicolaou and Motter [100] proposed a concept of mechanical metamaterials with negative compressibility, where the system moves from a stable state to a metastable state due to the applied loading. A similar concept of bistability was observed in origami- and kirigami-inspired metamaterials [45, 104, 131], where energy is stored in deformed elements. Another example of systems with negative compressibility are 3D metamaterials with cubic cells composed of hollow sealed functional elements, proposed by Qu et al. [114]. Materials with negative compressibility could be used as artificial muscles, actuators, pressure sensors or sensitive instruments in all applications where compression needs to be avoided (e.g. in deep ocean) [130, 145].

An interesting property that can also be achieved in mechanical metamaterials is negative stiffness [15, 145]. While traditional systems tend to deform consistently with the load direction, systems with negative stiffness deform in the direction opposite to the applied force. Therefore, they exhibit much larger deformations than the positive stiffness materials. They are, however, unstable and need to be constrained.

Lakes [79] proposed to use a combination of negative and positive stiffness materials to obtain unique properties, such as extremely high damping coefficients. Another example includes systems with buckled states, such as a simple planar beam buckled under compression, which was described by Klatt and Haberman [76]. The initial unbuckled state of the beam is unstable, which means that any perturbation will move the beam from its equilibrium. If a lateral constraint is applied and then removed, the beam will immediately take a new shape and this snap-through behaviour is the hallmark of the negative stiffness systems [145]. Negative stiffness materials have a wide range of potential applications, such as: composites with high damping coefficients, acoustics cloaking, vibration protection systems, seismic protection, etc. [145].

Based on the above literature study, it can be noticed that various extraordinary properties can be identified in mechanical metamaterials, including tensegrity-based ones. The following section further develops selected issues of extremal behaviour of metamaterials and metastructures, with a focus on adaptive systems.

2.4 ADAPTIVE SYSTEMS WITH EXTREMAL MECHANICAL PROPERTIES

In recent years, an increased interest in adaptive metamaterials and structures has been observed [48, 73, 83, 136]. These include active, responsive, programmable and deployable systems, which use such phenomena as large deformations and instability in order to enhance favourable mechanical properties of the systems. Below, three categories of adaptive systems are discussed: programmable metamaterials, origami- and kirigami-inspired systems and tensegrity-based metamaterials and lattices. All of the discussed systems exhibit various extraordinary behaviours, such as the extremal mechanical properties described in Section 2.2 or unusual properties presented in Section 2.3.

2.4.1 PROGRAMMABLE METAMATERIALS

The main idea behind the concept of adaptive metamaterials is the possibility of controlling the instability modes and thus, activating specific modes of instability when required [145]. Multistable metamaterials should be understood as systems that have more than one equilibrium configuration and the ability to switch between these states when a proper stimulus is applied [75]. For regulation of the instability behaviour of such materials, 2D and 3D cellular soft solids are used [80, 108, 122].

The idea of active and programmable mechanical metamaterials has initially emerged from the study on instability regions of soft and patterned materials [145]. Silva et al. [123] introduced a concept of metamaterial analog computing, based on blocks that can perform mathematical operations on the profile of a propagating wave. Such programmability level is hard to achieve in mechanical metamaterials, however, several attempts have been made.

Florijn et al. [48] created laterally confined mechanical metamaterials with programmable response to uniaxial compression. The developed systems exhibit monotonic, non-monotonic and hysteretic behaviour, which arises from a broken rotational symmetry. Kazemi and Norato [73] studied programmable lattice materials with struts that can be activated and deactivated by actuation mechanisms. They proposed a topology optimisation method for the design of such systems. Galea et al. [53] introduced a novel magneto-mechanical metamaterial that can change its linear dimensions when subjected to a uniform external magnetic field.

2.4.2 ORIGAMI-INSPIRED METAMATERIALS AND STRUCTURES

An interesting group of adaptive systems are origami- and kirigami-inspired mechanical metamaterials and lattices [47,83,120,136,147], which were already mentioned in Section 2.1. The term *origami* originates from the Japanese words: *ori* meaning *to fold*, and *kami* meaning *paper*, and it stands for an ancient art of paper folding. Kirigami is a variation of origami, where the paper is not only folded, but also cut—*kiri* means *to cut*. Origami- and kirigami-inspired engineering systems are based on plates, which are connected by compliant hinges to form various geometries with adjustable properties. Deformation mechanisms found in those systems result in extraordinary mechanical properties, such as multistability, large deformation, programmable stiffness, or negative Poisson's ratio [130].

One of the most popular origami patterns, which is used to create mechanical metamaterials as well as macro-scale lattices aimed at civil engineering applications, is the Miura-ori pattern. Schenk and Guest [120] proposed two folded systems based on this pattern: a folded shell structure and a cellular metamaterial obtained by stacking the individual folded layers. The authors showed that the folded shell structure exhibits a negative Poisson's ratio for in-plane deformations and, at the same time, a positive Poisson's ratio for out-of-plane deformations.

Although the Miura-ori pattern is most popular in engineering applications, many other origami-inspired systems have also been investigated. Filipov et al. [47] described origami tubes, which can be assembled into stiff, adaptive structures and metamaterials. They proved that the use of zipper-coupled origami tubes can increase the stiffness of the structure by two orders of magnitude. The developed origami systems can be applied in various scales, in deployable structures as well as in metamaterials that can be deployed, stiffened and tuned. Overvelde et al. [104] proposed an origami-based mechanical metamaterial with tunable shape, volume, and stiffness (see Figure 2.1d). They proved analytically and numerically that the adaptive metamaterial has three degrees of freedom, which allowed them to obtain various shapes through embedded actuation.

Wang and Wang [139] discussed kirigami-based metastructures with the focus on mechanical programmability and energy harvesting. They developed and validated a theoretical model that makes it possible to analyse certain mechanical properties of kirigami-inspired systems, such as flexibility, critical buckling strain, maximum tensile strain, elastic stretchability, and stiffness. Lee et al. [83] proposed a new type of compliant curved-crease origami-inspired metamaterials with a programmable force–displacement response. They showed that the parameters of the unit cell can be used to generate various response shapes. Moreover, they developed a new analytical method for the rapid prediction of the system response, which is based on the metamaterial geometric parameters.

2.4.3 TENSEGRITY-INSPIRED METAMATERIALS AND LATTICES

Apart from origami- and kirigami-based adaptive systems, tensegrity-inspired metamaterials and lattices deserve special attention. Tensegrities are pin-jointed systems consisting of struts (compressed elements) and cables (members in tension), which form a statically indeterminate structure in stable equilibrium [28, 98]. The most characteristic feature of these systems are infinitesimal mechanisms balanced with self-stress states [28, 78], which make it possible to apply tensegrities in adaptive systems with extremal mechanical properties. As tensegrity systems are the main focus of this study, they are described in detail in Chapter 3 and further developed in consecutive chapters. However, for the purpose of completeness, a state-of-the-art review on adaptive tensegrity-inspired mechanical metamaterials and lattices is presented here.

The concept of tensegrity metamaterials was first introduced by Fraternali et al. [51] for the dynamics of the chain of tensegrity prisms and then, further developed by the authors in [11, 44]. Fabbrocino et al. [44] proposed a numerical method for analysing the nonlinear wave-dynamics of tensegrity systems subject to impulsive compressive loading. They showed that tensegrity lattices have a great potential in engineering applications, for example, as sensors and actuators in structural health monitoring and damage detection systems. A similar concept was analysed by Wang et al. in [138]. The authors focussed on lightweight metastructures that consist of prismatic tensegrity modules and, using the developed theoretical model, they studied tunable properties of the proposed systems. They proved that it is possible to obtain tensegrity lattices with tunable stiffness and wave propagation, either by using external control of geometrically nonlinear deformation or by adjusting prestressing forces in structural members.

De Tommasi et al. [37] focussed on the mass and morphological optimisation of tensegrity-based mechanical metamaterials. They proposed a prototypical model of a compressed slab tessellated into triangular, square, and hexagonal cells. Modano et al. [97] proposed tensegrity metamaterials based on novel unit cells. They analysed various configurations of the lattices and showed that the described systems exhibit a large number of infinitesimal mechanisms, which can be effectively stabilised with self-stress forces. Geometrically nonlinear behaviour of uniformly compressed tensegrity prisms was recognised by Fraternali et al. [50]. The authors developed elastic and rigid-elastic models that allowed them to predict the mechanical behaviour of tensegrities, including extremal stiffening and softening response of the systems, which could be applied in tensegrity lattices and innovative metamaterials.

An interesting method to construct three-dimensional tensegrity lattices from truncated octahedron elementary cells was proposed and discussed by Rimoli and Pal [117] and extended for phase transition by Salahshoor et al. [118]. Various automatically assembled tensegrity lattices were proposed by Zhang et al. [148] for large scale structures. Zhang et al. [149] and Ma et al. [90] introduced metal rubber into the struts of tensegrity prisms. Both the theoretical and experimental data showed a significant improvement of energy absorption and tunable dynamic properties, what is a big advantage as far as efficient adaptive mechanical metamaterials are concerned. Bauer et al. [18] discussed a concept of delocalised deformation in failure-resistant tensegrity metamaterials and structures. The proposed systems exhibited a significant enhancement of deformability and energy absorption capability without failure and thus, could be applied in impact protection systems as well as adaptive load-bearing structures.

Liu et al. [87] developed a systematic design approach for generation of tensegrity metamaterials with various inner architectures. The authors demonstrated that mechanical metamaterials based on tensegrity cells offer tunable mechanical properties (effective elastic moduli, Poisson's ratio) by adjusting the level of prestress, which makes them an excellent solution for programmable systems. Adjustable mechanical properties of tensegrity lattices were also discussed by Al Sabouni-Zawadzka and Gilewski in [5, 8, 9]. In [8] the authors studied smart properties as well as a negative Poisson's ratio of an orthotropic metamaterial based on the simplex tensegrity pattern. Extremal properties of regular tensegrity unit cells in 3D lattice metamaterials were analysed in [5], and in [9] the authors proved that the metamaterials built from such modules can also exhibit extremal mechanical properties.

Above in Sections 2.4.1–2.4.3, three groups of adaptive systems were discussed: programmable metamaterials, origami- and kirigami-inspired systems, and tensegrity-based metamaterials and structures. It was demonstrated that such systems differ from traditional structural materials or structures, as they can exhibit extraordinary properties, such as for example extremal mechanical behaviour described in Section 2.2. Adaptive systems are not the only metamaterials or metastructures with extremal mechanical properties, however, they have attracted much attention of the scientists in recent years, as they have a wide range of potential applications in engineering.

3 Tensegrity—Smart Structures with Unusual Mechanical Properties

3.1 SMART STRUCTURES

The concept of smart structures emerged as a distinct field of applied sciences in the 1980s, when due to the ongoing technological progress, researchers started to work together on the development of smart structural systems [12]. At first, all research programs carried out within this area were funded by the government. In the early 1990s private companies began to invest money, providing necessary funding and application possibilities. The cooperation between government research organisations, scientists and companies resulted in several multi-year programs dedicated to the development of smart products and their implementation.

Smart systems have a wide range of applications in various areas [3,46,86]: from aerospace and space engineering, through automotive industry, robotics and biomedical engineering, up to civil engineering. The last field of implementation is relatively new and for a long time, it has been dominated by other areas.

In aerospace engineering the smart technology is used for damage detection, shape and vibration control and a variety of on-board systems. In automotive industry, smart systems ensure a passenger comfort and a health monitoring of vehicles through an active vibration control and acoustic damping. Smart products are also used in a widely understood industry: for machine chatter control in manufacturing, vibration control in mining machinery, etc. Another very important field of application is biomedical engineering, where a fast development of smart surgical tools can be observed.

As far as civil engineering applications are concerned, several examples of studies on smart structures can be found in the literature. Chang and Spencer [33] carried out research on a two-storey model building made of steel, equipped with six low-friction pendulous bearings and three low-force hydraulic actuators. The aim of the control was to limit base displacements, reducing at the same time floor accelerations. Cazzulani et al. [30, 31] investigated a carbon fibre structure, which was controlled by an active damping and dissipation of mechanical energy of the structure, aimed at stress and vibration reduction. Lu et al. [88] tested a control system used for a continuous structural monitoring and seismic response control of the Civil Engineering research building at the National Taiwan University.

DOI: 10.1201/9781003343202-3

Most of the studies on smart structures, including the ones mentioned above, focus on advanced technologies and control systems, which are installed in traditional structures. There are, however, systems that are more prone to structural control than typical structures applied in civil engineering. Examples of such systems were already presented in Section 2.4, where adaptive tensegrity-based metamaterials and lattices were discussed. Similar systems can be applied in a bigger scale to form structural elements such as columns, plates, beams, or even whole civil engineering structures, such as bridges or towers.

A group of scientists from École Polytechnique Fédérale de Lausanne investigated a full-scale smart tensegrity structure [1, 19], which was controlled by an active modification of the self-stress state between struts and cables. When it came to local damage, the actively controlled smart system compensated a broken element, satisfying the serviceability criteria. Rhode-Barbarigos [116] presented a concept of a deployable tensegrity footbridge. The prototype of the structure, constructed by the scientist of the Swiss Federal Institute of Technology, consisted of four pentagonal tensegrity modules with a total span of 16 m and was equipped with an advanced active control system. Tibert [133] presented a smart deployable tensegrity mast that was based on the system developed by Snelson [126], constructed from overlapped regular simplex modules. Another structure of this type was proposed by Gonzales et al. in [60], where the authors presented an algorithm for the reconfiguration of a two-stage tensegrity mast. Zawadzki and Al Sabouni-Zawadzka [10, 146] analysed deployable tensegrity columns with structural properties controlled by the self-stress modification.

3.1.1 DEFINITIONS

The definition of smart structures has always been a disputable issue. Authors [2, 56, 129, 135] define and classify smart structures in different ways. The discrepancies result from distinct ways of perceiving behaviour of such structures and methods of their analyses. Some examine a whole structure, while others focus on a specific part of it—a single structural element.

Another reason for such a variety of existing definitions is the fact that the word *smart* itself has various meanings. According to the dictionary, its original significance was *stinging, sharp*. The present meaning of the word *smart—clever, intelligent*—has taken over from its original definition the element of quick energetic movement and sharp thought. This original meaning characterises perfectly the idea of smart structures. In relation to structures, the word *smart* means: capable of acting in a quick way and making corrections that resemble human decisions, particularly in response to changeable conditions. The present significance—*intelligent*—is also applied to structures, but it is not fully adequate. Intelligence is a human feature and should be reserved for humans. Although the classification distinguishes a group of very smart structures, calling them intelligent, it does not mean that those structures possess intelligence. The only attribute, which makes them resemble humans, is the ability to learn.

It should be noted that there are various types of *smartness*. Let us consider a smart building as an example. For its user, the smart technology will be associated with high-tech equipment, such as electronics controlling the ambient environment, air conditioning systems, lighting and alarm installations. For an engineer designing the structure, it will probably mean that the building is equipped with a smart monitoring system.

There are two concepts that can be distinguished within the field of smart systems: smart structures and smart materials.

Smart structures can be defined as structural systems with the ability to sense and respond in an adaptive way to changes in the surrounding environment. This feature distinguishes them from the conventional ones. Whereas the main purpose of traditional structures is to carry loads and provide safety for the users, the smart ones additionally adapt in a pre-designed manner to the functional need, by modifying their shape, stiffness or damping characteristics in order to minimise deflection and possible damage.

Smart materials are materials which are able to convert one form of energy (mechanical, magnetic, electrical, etc.) into another, in a reversible and repeatable process. They are capable of sensing changes in the environmental conditions, responding to them in a predetermined manner, in an appropriate time and returning to their original shape as soon as the stimulus is removed. Smart materials are often used in actuation systems of smart structures, stimulating them to adapt to variable conditions.

Smart systems should be understood as complex structural systems that can be composed of smart materials, smart structures or both, and additionally equipped with expert data processing. Smart systems ensure that during normal conditions the structure carries all the loads without any help of smart components and, at the same time, it uses specific actuation systems to tackle abnormal load cases. Sometimes, the term *smart system* is used interchangeably with the term *smart structure*, regarding the structure as a whole structural system.

Adaptive metamaterials and lattices that were discussed in Section 2.4 should be classified as smart structures rather than smart materials, as they are structural systems with tunable properties, which are able to adapt to the current conditions or desirable state thanks to their unique mechanical properties, and not because of their chemical composition.

3.1.2 CLASSIFICATION

Smart structures can be divided in three main categories, depending on the applied control system:

- passively controlled structures,
- actively controlled structures,
- very smart (intelligent) structures.

Passively controlled structures are systems containing sensors used for structural health monitoring (SHM) and actuators used for the alteration of system characteristics. They are able to respond to external stimuli in a controlled manner, without the use of any electronic control devices or feedback systems.

Actively controlled structures include sensors and actuators connected together in an integrated smart system, which controls both characteristics of the structure and its functionality. Actively controlled structures utilise feedback systems to improve sensing and reaction processes.

Very smart (intelligent) structures are capable of learning and memorising past control commands. They use nonlinear properties of sensors, actuators and feedback systems in order to adapt to the new conditions. Intelligent structures often contain an artificial neural network (ANN).

Actively controlled systems consist of three key elements [3]: sensors, actuators and a control unit (Figure 3.1). Sensors are elements that detect changes in the environment, record the structural response (stress, strain, etc.) and generate proper signals, which are then sent to the control unit. The control unit is an element responsible for data analysis. It gathers all the information received from sensing devices, processes them and, based on the given algorithm, reaches the conclusion about further action. If a specific response is required, the control unit sends a signal to the proper actuator. Actuators are elements responsible for the change of structural response. They change properties of the structure by applying a signal that was computed by the control unit. This makes it possible to reduce the structural damage and avoid the catastrophic global collapse.

These elementary components work together using signal transferring devices which transmit: first collected data and then produced control commands between separate parts of the system. As far as sensors and actuators are concerned, there are numerous types of devices used [31, 46, 99]. The most popular are Fibre Bragg Grating sensors (FBG) and piezoelectric actuators (PZT). Other commonly utilised actuators are shape-memory alloys (SMAs) and, recently, mechanical metamaterials described in Chapter 2.

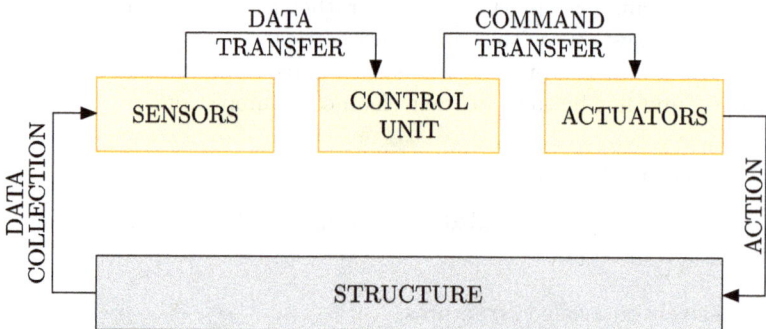

Figure 3.1 Components of a smart actively controlled system.

The idea of smart systems introduced above is not always related to metamaterials or metastructures that were described in Chapter 2. There are systems with extremal or other extraordinary mechanical properties, which cannot be classified as smart materials or structures. However, tensegrity-based systems, both the ones in a material scale and bigger structural systems, exhibit some unique properties making them perfect for active control. This is why the analysis of tensegrity systems should not be carried out in isolation from the concept of smart structures.

3.2 TENSEGRITY

Among smart structural systems with unusual mechanical properties, tensegrity concept is of particular interest in engineering sciences. The term *tensegrity* was first introduced by Buckminster Fuller (see Skelton and Oliveira [125] for historical details) as a combination of two words: *tension* and *integrity*.

Several definitions of this concept can be found in the literature [72, 98, 125, 133]. First definitions appeared in patent documentations by Fuller [52] and Snelson [126]. Both authors highlighted the occurrence of a discontinuous set of compressed elements inside a continuous set of elements in tension. Emmerich [40] was the first to mention a very important property of tensegrity structures—the occurrence of self-equilibrated system of normal forces, called self-stress. Motro in [98] presented a definition that was based on first tensegrity patents, stating that tensegrities are spatial grid systems in self-stress state. In his definition, he also mentioned that there should be only one compressed element per node, which significantly limited the class of tensegrity systems to a narrow set of structures. Miura and Pellegrino (see [133]), on the other hand, presented a definition that was based on mechanics of structures, clearly defining the requirement for the occurrence of infinitesimal mechanisms. A slighlty different approach was proposed by Skelton and Oliveira in [125]. They defined a tensegrity configuration as a set of rigid bodies which become tensegrity after connecting them with elements in tension. Additionally, they mentioned the possibility of joining more than one compressed member in one node. According to the authors, the number of compressed elements connected in one node defined the tensegrity class. Kasprzak [72] presented another idea for the tensegrity definition. He defined two types of systems: pure tensegrities and tensegrity-like structures. He emphasised the role of infinitesimal mechanism balanced with self-stress states in pure tensegrities.

In this study, tensegrity structures are defined as pin-jointed systems with a particular configuration of cables and struts that form a statically indeterminate structure in stable equilibrium. Tensegrities consist of a discontinuous set of compressed elements (struts) inside a continuous set of members in tension (cables), which have no compression stiffness. A unique property of tensegrity structures are infinitesimal mechanisms, which are balanced with self-stress states [7, 28, 78, 115]. The occurrence of a self-stress state in the structure indicates that there is a certain set of internal forces in structural members, which are independent of external loading and boundary conditions because they are in self-equilibrium. Moreover, it is assumed that more than one strut can be connected in each node and this number defines the tensegrity class (as in Williamson and Skelton [141]): a tensegrity system of class n is a structure in which at most n struts are connected to any node.

Tensegrity systems have many advantages, such as:

- high stiffness-to-mass ratio,
- free geometry shaping,
- controllability,
- deployability,
- reliability,
- smart properties,
- extremal mechanical properties.

High stiffness-to-mass ratio is a very important feature from the practical point of view, as it creates the possibility to construct lightweight systems with a high load-bearing capacity and a low material usage. Free geometry shaping results from various forms of available tensegrity modules, the possibility of joining these basic structures and of creating new, non-modular systems. It is especially important for architects, as it allows them to design structures with almost any shape and function. Controllability of tensegrity systems consists in the possibility of controlling their structural properties by adjusting the level of self-stress (this property is discussed in detail in Section 3.3). Deployability of tensegrities was discussed thoroughly by Al Sabouni-Zawadzka and Zawadzki in [10, 146] and this feature refers to systems, which have the ability of changing their shape. Reliability should be understood as the possibility to be used in various conditions, to undergo self-repair in case of damage (see Section 3.3), and to maintain a sufficient level of load-bearing capacity even in extreme loading conditions.

The last two mentioned advantages—smart and extremal properties of tensegrity structures—are discussed in detail in this study. Inherent features of tensegrities are described in Section 3.3. They allow us to classify tensegrities as smart structures (according to the definition given in Section 3.1.1) and to apply them in situations, where certain structural characteristics need to be adjusted during the operation of the structure, for example, bridge structures with adjustable stiffness, tensegrity plates with adjustable damping characteristics, deployable systems with changing geometry. Extremal properties, which are the main focus of this study, are in many ways related to smart features. They are analysed in Chapter 5 and imply that the system is extremely stiff under certain stresses and extremely compliant in other orthogonal cases of stresses. Moreover, some of extremal systems exhibit unusual mechanical properties, such as for example a negative Poisson's ratio. This feature of tensegrity structures can be regarded as both advantage and disadvantage, as in some typical engineering applications the structures should not exhibit extremal behaviour and their properties should maintain constant. At the same time, however, there are applications which require adjustable properties, such as mechanical metamaterials, elements used for absorption of shocks and seismic vibrations, deployable systems etc.

To main disadvantages of tensegrity systems belong:

- risk of struts congestion,
- fabrication complexity,
- problems with joints design,
- complex geometry,
- lack of materials with sufficient strength,
- too high compliance in certain directions.

The first mentioned disadvantage—a risk of struts congestion—becomes a problem in large designs, where due to the complex geometry, the compressed members start running into each other. This problem, however, can be quite easily solved by careful geometry planning, applying form-finding techniques or using modular structures. A serious disadvantage is the fabrication complexity. Construction of real large-scale tensegrity structures is a big challenge as they are not very often used and therefore, the fabrication technology has not been well developed. The next two disadvantages are related to this issue. Tensegrities have a complex geometry, which requires special design, analysis and construction technology and, what is the biggest challenge—design and fabrication of joints. In tensegrities, many structural members coming from different directions meet together in one node and the joints should allow the elements to rotate freely without generating bending moments. In many systems, such as deployable structures, these free rotations are particularly important, as their geometry changes drastically during the deployment. Another disadvantage is a lack of materials with sufficient compressive or tensile strength. Using the commonly available materials, it is not always possible to increase the values of prestressing forces in the structures to the level that would guarantee the occurrence of extremal properties. This, however, will probably change in the future, as more and more new materials with special properties are being developed. The last of the listed disadvantages is a high compliance of certain tensegrity systems in particular directions. This issue was already discussed above, while describing the advantages of tensegrities—extremal properties are not always desirable in many typical civil engineering applications.

3.3 INHERENT PROPERTIES OF TENSEGRITY

Smart systems include sensors, actuators and control mechanisms, which are integrated in one coherent system that becomes an integral part of the structure. However, neither the external devices installed on the structure nor the applied intelligent technologies make the structure smart. The structure itself should have some special properties, hereinafter referred to as inherent, which make it possible to control it without any external interference.

To inherent properties of smart structures [7] belong:

- self-control,
- self-diagnosis,
- self-repair,
- self-adjustment (active control).

As it was proved by Al Sabouni-Zawadzka and Gilewski [7], tensegrity structures have all of the above mentioned properties, which they owe to infinitesimal mechanisms balanced with self-stress states. In the following sections, several examples of analyses of tensegrity modules and multi-module structures are presented, in order to demonstrate that it is possible to control their properties by adjusting the prestressing forces.

Self-control of tensegrity systems consists in self-stiffening of the structures under the applied load that causes displacements consistent with the infinitesimal mechanism mode. External loading acts similarly to the self-stress—it eliminates singularity of the problem, additionally prestresses the structure and stiffens it.

Self-diagnosis relates to the possibility of damage detection and identification by measuring internal forces in active members. Damage of one structural element affects the distribution and level of self-stress in the whole structure.

Self-repair of tensegrity structures is realised by adjusting self-stress forces. A proper change of prestressing level can compensate the damaged element and restore the values of structural displacements from before damage.

Self-adjustment (active control) of tensegrity systems is a result of self-stress states. Prestressing of the whole structure as well as its selected part results in stiffening of the system and reduction of its displacements. Therefore, active control of tensegrity can be realised by adjusting the level of self-stress in only one selected part of the structure.

3.3.1 SELF-CONTROL

Due to the infinitesimal mechanisms that occur in tensegrity systems, it is possible to control their properties by adjusting the values of prestressing forces in structural members. However, the adjustment of self-stress is not the only method of changing properties of these structures. Thanks to their inherent smart features, they are capable of self-control—they stiffen under the applied external load, which causes displacements that are consistent with the infinitesimal mechanism modes. In order to examine this unique property of tensegrity, a geometrically nonlinear analysis has to be used. Analysis according to the 2nd order theory, which includes self-stress but neglects geometrical nonlinearities, makes it possible to identify infinitesimal mechanisms and determine self-stress states of the structure, but does not take into account self-stiffening of the structure. Therefore, the analyses presented in this section are performed according to the geometrically nonlinear theory.

Self-control can be observed in all types of tensegrity systems, simple as well as complex ones. Below, an example of a single tensegrity module is presented—a four-strut simplex. The analysed structure consists of cables with a diameter of 20 mm, made of steel S460N, with a load-carrying capacity of 110.2 kN, and struts—compression rods with an outer diameter of 76.1 mm and a wall thickness of 2.9 mm, made of steel S355, with a load-carrying capacity of 160 kN. The calculations were performed in SOFiSTiK software, using the finite element method (FEM) [17, 35, 68], within the geometrically nonlinear theory.

The values of self-stress forces are presented in Figure 3.2, where S_0 is a multiplier of the prestressing force and is used to indicate the level of self-stress.

The module (Figure 3.3) was loaded with a concentrated force F applied to the upper node of the structure. Two types of parameters were measured in the analysis: a horizontal displacement u of the loaded node and forces in two structural members—cable N_1 and strut N_2.

The structure was analysed twice, using two levels of loading: $F = 10$ kN and $F = 20$ kN. Results of the performed analysis are presented in Figure 3.4. It can be noticed that the external load, similarly to self-stress, leads to the stiffening and additional prestressing of the structure. Its influence on the global stiffness of the system depends on the ratio between prestressing forces in structural members and the forces caused by self-stiffening under the applied load. Comparison of internal forces in cable 1 and in strut 2 with the prestressing forces (calculated according to Figure 3.2) leads to the conclusion that self-stress forces have bigger influence on the global stiffness of the structure when the values of external loads F are small. When F increases, external loading starts to play an important role in self-stiffening of the system.

It can be concluded that properties of tensegrity structures depend not only on self-stress. Tensegrities have the ability of self-control, which is realised

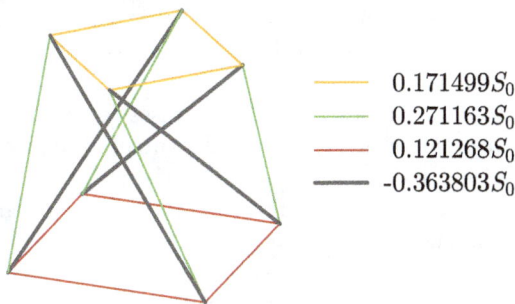

$$
\begin{array}{l}
\text{——} \quad 0.171499S_0 \\
\text{——} \quad 0.271163S_0 \\
\text{——} \quad 0.121268S_0 \\
\text{——} \quad \text{-}0.363803S_0
\end{array}
$$

Figure 3.2 Self-stress state in the four-strut simplex.

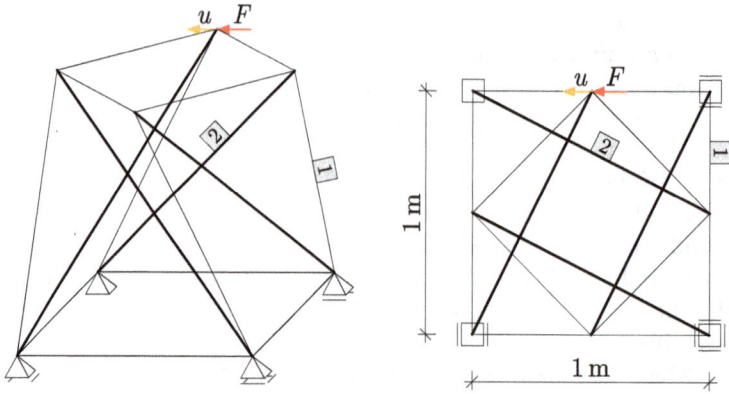

Figure 3.3 Geometry and loading of the four-strut simplex.

by self-stiffening of the structure under the applied external load that acts in the direction that is consistent with the infinitesimal mechanism mode. This feature is particularly important in non-typical loading cases, which are hard to foresee.

3.3.2 SELF-DIAGNOSIS

One of the aims of currently conducted research on structural health monitoring (SHM) is the development of systems, which would allow for the detection of damage, based on the analysis of data obtained from SHM [23, 29]. Such a structural diagnosis is based on various smart technologies that include a series of external devices installed on the structure and is independent from structural properties. In the case of smart structures, the diagnosis of damages should be possible thanks to the special internal properties of the structure, not the smart devices with which the structure is equipped. Tensegrity structures are a good example. The analyses presented in the previous section proved that tensegrities are capable of self-control due to the infinitesimal mechanisms, which are balanced with self-stress states. This unique feature may be used not only for structural control but also in the case of damage detection systems.

Failure of a random structural member causes redistribution of nodal displacements and internal forces in cables and struts. In the case of multi-module structures, failure of one element eliminates self-stress within the module that contains it. Therefore, identification of the damaged area can occur immediately. Moreover, damage detection in tensegrity structures can be realised by measuring internal forces only in active cables and struts, as the failure of one random structural member causes a redistribution of prestressing forces within the whole structure. The analysis presented in the next section shows how the failure of one structural element influences the behaviour of the whole tensegrity structure.

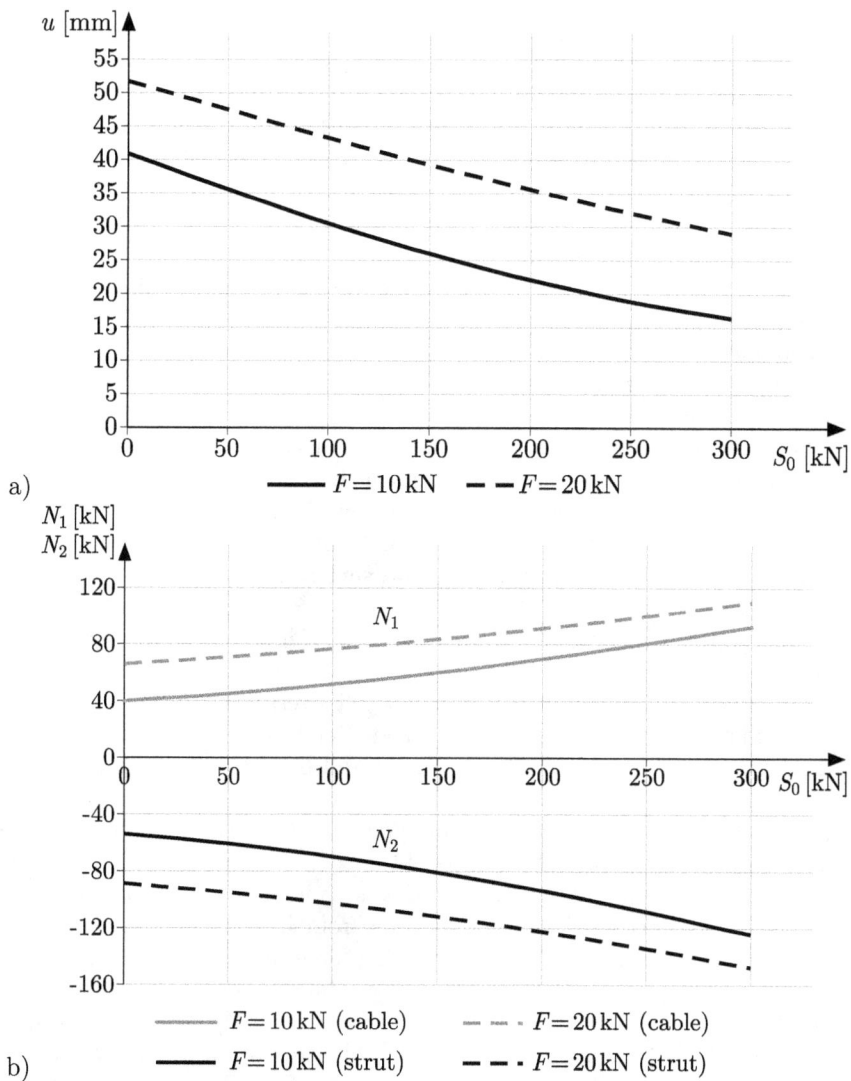

Figure 3.4 Influence of self-stress on: a) nodal displacement; b) normal forces in cable 1 and strut 2; for two levels of loading.

3.3.3 SELF-REPAIR

In order to prove that tensegrity systems are capable of self-repair, an analysis of a tensegrity plate-like structure was performed. In [7] the ability of self-repair was shown using an example of a tensegrity structure constructed from three-strut simplex modules. Here, a multi-module structure based on four-strut simplex units (Figure 3.5) is analysed. The modules are rotated clockwise and counter-clockwise and connected together in nodes. The structure is loaded with concentrated forces $F = 1$ kN applied vertically to the nodes of the upper surface. The same materials, dimensions of the cross-sections and values of self-stress were used as in the single module described in the previous section.

The aim of the analysis was to examine behaviour of the structure in case of a member loss and to verify the possibility of self-repair. The failure was modelled by removing one of the cables from the system—marked with a dashed line in Figure 3.5. The calculations were conducted in SOFiSTiK software, with the use of the finite element method, according to the geometrically nonlinear theory.

Within the analysis, a possibility of structural repair through an adjustment of self-stress in one selected module of the structure was verified. Such a possibility would be particularly valuable from the practical point of view—it would allow for the reduction of the number of actuators installed on active structural members. As it is proved below, such a repair is possible in the case of the analysed system.

The scope of the analysis was a vertical displacement of the central node of the upper surface of the plate-like structure, marked with letter A in Figure 3.5. For the loading $F = 1$ kN and self-stress level $S_0 = 100$ kN, the displacement $v_A = 65.5$ mm was obtained. Removal of the cable resulted in changes in the structure of modules no. 2 and 6 (Figure 3.5), so they could no longer be prestressed after the member loss. Self-stress on the level of $S_0 = 100$ kN was maintained in six undamaged modules. After the simulated damage the value $v_A = 92.2$ mm was obtained.

Afterwards, repair of the structure was simulated, aimed at the reduction of the displacement v_A to the value from before damage. The repair was performed by adjusting the prestressing forces in single undamaged modules. Each time the level of self-stress within the considered module was changed, and other modules remained prestressed with the force multiplier $S_0 = 100$ kN. Figure 3.6 shows levels of self-stress obtained for each module, that allowed for regaining the initial value of v_A. The smallest values of prestressing forces that led to self-repair of the structure were obtained for module no. 5.

Comparison of two tensegrity plate-like structures—the one presented here and the one analysed in [7]—proves that each tensegrity system is different.

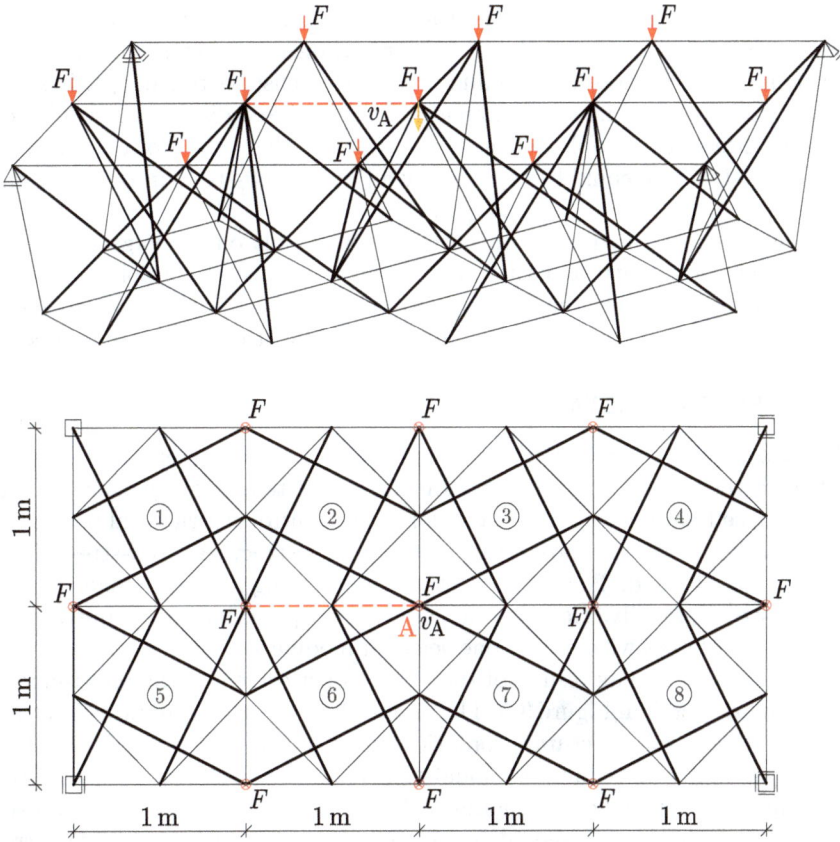

Figure 3.5 Multi-module tensegrity structure with member loss.

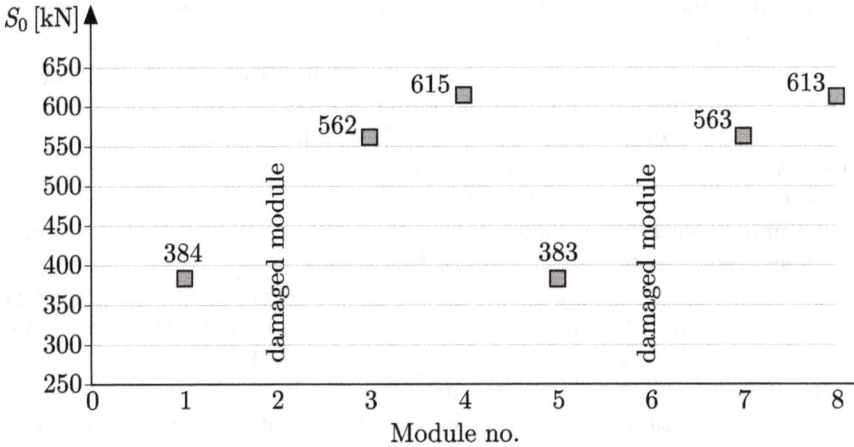

Figure 3.6 Levels of self-stress required for self-repair.

Although both structures are constructed in a similar way and both are based on simplex modules, they exhibit different stiffness characteristics. The four-strut simplex system is less stiff and much more sensitive to damage. Removal of the cable resulted in a 40% increase of the analysed displacement, whereas for the three-strut simplex structure it was less than 8%. However, in both cases the structure could be repaired by adjusting self-stress in the remaining modules.

The example presented in this section covers only one damage scenario in one specific tensegrity structure and does not exhaust the problem of structural repair. However, the analysis proved that tensegrities are capable of self-repair that is performed through the adjustment of prestressing forces.

3.3.4   SELF-ADJUSTMENT

Active control of smart structures consists in the adjustment of their behaviour by using their inherent properties. In the case of tensegrity systems it is the ability of self-adjustment that allows for active control of structural properties.

In order to control properties of tensegrity structures, self-stress forces need to be adjusted. The most effective way of doing it is to change the level of self-stress globally. However, it can also be performed by adjusting the prestressing forces within only one selected module.

Figure 3.7 presents results of the analysis performed on the multi-module structure depicted in Figure 3.5. The aim of the analysis was to prove that it is possible to control structural properties—in this case: increase stiffness of the structure and thus, reduce the nodal displacement v_A—by changing the prestressing forces in one of the modules. The calculations were performed using the 2nd order theory. It should be noticed that the adjustment of self-stress in the single module did not reduce the displacements to the same degree as the self-stress applied to the whole structure did. However, it eliminated the infinitesimal mechanism and decreased the value of the considered displacement significantly. Depending on the module, the influence of self-stress on structural stiffness varied. The biggest reduction of the displacement v_A was obtained by adjusting the forces in module no. 5.

Figure 3.7 shows how the self-stress level influences the displacement of node A. The structural control was performed in two ways: by adjusting the prestressing forces in the whole structure and by changing the level of self-stress in only one module—module no. 5. In the second scenario the remaining modules were prestressed with forces $S_0 = 20$ kN.

The presented example proves that the active control of tensegrities can be realised by adjusting the level of self-stress in the whole structure, as well as in only one selected part of it. Prestressing of both the whole system and its part leads to stiffening of the structure and reduction of its displacements.

Above in Sections 3.3.1–3.3.4 four very important properties of tensegrity systems were discussed. It was demonstrated that tensegrities exhibit a series of unique mechanical properties, which make them prone to structural

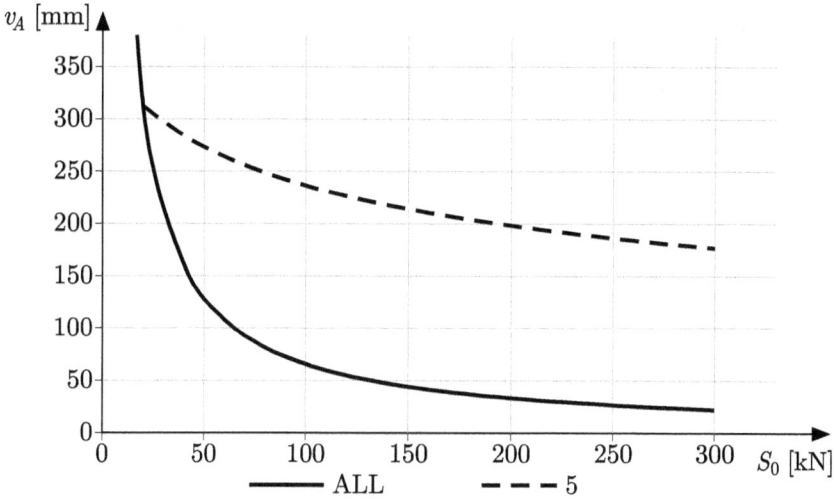

Figure 3.7   Influence of self-stress in module no. 5 and in all modules on the vertical displacement of node A.

control, independently from external devices used in typical smart systems. Although the examples presented above are not directly related to extremal mechanical behaviour, which was discussed in Chapter 2, they are essential to understand the nature of tensegrity structures that is inextricably linked to the occurrence of infinitesimal mechanisms balanced with self-stress—the same characteristics that are responsible for extremal behaviour of these systems.

4 Analysis of Tensegrity Systems

4.1 DISCRETE MODELS

Tensegrity structures can be relatively easily described by means of discrete models, either by using the finite element method (FEM) [17, 152] or by directly formulating the tasks algebraically [85, 110, 111]. Both techniques are formally precise and make it possible to include in the description an influence of self-stress on structural response. The finite element formalism is algorithmically easy and is based on standard FEM steps, from the analysis of a separate element, through globalisation, to considering boundary conditions. The algebraic direct formulation is mathematically elegant, but requires a formulation of global matrices at the beginning of the process. The equivalence of both techniques was demonstrated by Pełczyński and Gilewski in [110].

From the mechanical point of view, tensegrity systems can be qualified as trusses and described using the following parameters (the parameters are defined for an unsupported truss):

$\mathbf{q}$—vector of nodal displacements (length m),
$\boldsymbol{\Delta}$—vector of extensions in bars (length n),
$\mathbf{N}$—vector of normal forces in bars (length n),

$$\mathbf{D} = \begin{bmatrix} \dfrac{E_1 A_1}{L_1} & 0 & \cdots & 0 \\ 0 & \dfrac{E_2 A_2}{L_2} & \cdots & 0 \\ \cdots & \cdots & \cdots & \cdots \\ 0 & 0 & \cdots & \dfrac{E_n A_n}{L_n} \end{bmatrix} \text{—elasticity matrix,}$$

$\mathbf{P}$—vector of nodal loads (length m),

where: E_e is a Young's modulus of the e-th element, A_e is a cross-sectional area of the e-th element, and L_e is the e-th element length, $e \in (1, n)$.

If tensegrity structures are considered, one can identify a self-equilibrated system of normal forces $S_1, S_2, \ldots, S_n$, called self-stress. Equations of equilibrium can be defined in actual configuration, taking into account their nonlinearity. Several additional parameters need to be defined for tensegrities:

$\boldsymbol{\omega}$—vector of rotations of bars (length n),

DOI: 10.1201/9781003343202-4

F—vector of additional generalised forces in bars (length n),

$$\mathbf{S} = \begin{bmatrix} \dfrac{S_1}{L_1} & 0 & \cdots & 0 \\ 0 & \dfrac{S_2}{L_2} & \cdots & 0 \\ \cdots & \cdots & \cdots & \cdots \\ 0 & 0 & \cdots & \dfrac{S_n}{L_n} \end{bmatrix}.$$

Relations between these parameters can be described as follows:

$$\boldsymbol{\Delta} = \mathbf{Bq}, \quad \boldsymbol{\omega} = \mathbf{Cq}, \tag{4.1}$$

$$\mathbf{N} = \mathbf{D}\boldsymbol{\Delta}, \quad \mathbf{F} = \mathbf{S}\boldsymbol{\omega}, \tag{4.2}$$

$$\mathbf{B}^{\mathrm{T}}\mathbf{N} + \mathbf{C}^{\mathrm{T}}\mathbf{F} = \mathbf{P}, \tag{4.3}$$

where:
B—compatibility matrix,
C—matrix of rotations.

Matrix **B** makes it possible to express extensions of bars using nodal displacements, matrix **C** represents rotations of bars. After the substitution of Eq. 4.1 and Eq. 4.2 into Eq. 4.3, a system of displacement-based equations is obtained:

$$\mathbf{Kq} = \mathbf{P}, \tag{4.4}$$

where:
$\mathbf{K} = \mathbf{K_L} + \mathbf{K_G}$,
$\mathbf{K_L} = \mathbf{B}^{\mathrm{T}}\mathbf{DB}$—linear stiffness matrix,
$\mathbf{K_G} = \mathbf{C}^{\mathrm{T}}\mathbf{SC}$—geometric stiffness matrix

Using the above symbols, the Maxwell's truss equation for a statically determinate 3D unsupported structure can be formulated as:

$$m - n - r = 0, \tag{4.5}$$

where:
m—number of degrees of freedom,
n—number of bars,
r—number of required constraints.

The above formulation can be generalised for structures which have a smaller number of bars, but are still able to carry loads in all directions

[27], such as tensegrity systems. Such structures have infinitesimal mechanisms balanced with self-stress states. An infinitesimal mechanism describes a local geometrical changeability within the infinitely small displacements, while a finite mechanism is related to any geometrical changeability of the structure.

The generalised Maxwell's equation for a 3D tensegrity truss has a form

$$m - n - r = f - s, \qquad (4.6)$$

where:
f—number of infinitesimal mechanisms,
s—number of self-stress states.

The identification of infinitesimal modes and distribution of self-equilibrated normal forces can be performed using the singular value decomposition (SVD) of the compatibility matrix $\mathbf{B}$ [85, 109, 111].

Matrix $\mathbf{B}$ can be determined directly from the geometry of the structure, but in the case of 3D tasks it is more practical to use the FEM formulation [17, 35, 68]. Algorithm for the determination of matrix $\mathbf{B}$ according to FEM is described below:

1. Determination of nodal coordinates.
2. Determination of lengths and direction cosines of bars.
3. Determination of generalised Boolean matrices for elements $\mathbf{C}_e$ ($e = 1, \ldots, n$), with the dimensions of $m \times 2$—allocation and transformation.
4. Determination of extensions of bars: $\mathbf{\Delta}_e = \mathbf{b}_e \mathbf{q}$, $\mathbf{b}_e = \mathbf{B}_e \mathbf{C}_e$, $\mathbf{B}_e = [-1 \quad 1]$.
5. Determination of matrix $\mathbf{B}$ from row matrices $\mathbf{b}_e$: $\mathbf{B} = \begin{bmatrix} \mathbf{b}_1 \\ \mathbf{b}_2 \\ \cdots \\ \mathbf{b}_n \end{bmatrix}$.

The linear stiffness matrix $\mathbf{K}_L$ can be determined from the expression $\mathbf{K}_L = \mathbf{B}^T \mathbf{E} \mathbf{B}$ or by using the FEM formalism—through the aggregation of linear stiffness matrices for single elements $\mathbf{K}_{Le}$, transformed from the local coordinate system into the global one by means of the generalised Boolean matrices $\mathbf{C}_e$ related to parameters $\mathbf{q}_{Le}$:

$$\mathbf{K}_{Le} = \frac{E_e A_e}{L_e} \begin{bmatrix} 1 & -1 \\ -1 & 1 \end{bmatrix}, \qquad \mathbf{q}_{Le} = \{u_1, u_2\}, \qquad (4.7)$$

$$\mathbf{K}_L = \sum_{e=1}^{n} \mathbf{C}_e^T \mathbf{K}_{Le} \mathbf{C}_e. \qquad (4.8)$$

The global matrix $\mathbf{K}_G$ can be determined similarly to $\mathbf{K}_L$, by using the Boolean matrices $\overline{\mathbf{C}}_e$ related to the parameters $\mathbf{q}_{Ge}$ (with the dimensions of

$m \times 6$) and the local element matrix $\mathbf{K}_{Ge}$:

$$\mathbf{K}_{Ge} = \frac{S_e}{L_e} \begin{bmatrix} 0 & 0 & 0 & 0 & 0 & 0 \\ 0 & 1 & 0 & 0 & -1 & 0 \\ 0 & 0 & 1 & 0 & 0 & -1 \\ 0 & 0 & 0 & 0 & 0 & 0 \\ 0 & -1 & 0 & 0 & 1 & 0 \\ 0 & 0 & -1 & 0 & 0 & 1 \end{bmatrix},$$

$$\mathbf{q}_{Ge} = \{u_1, v_1, w_1, u_2, v_2, w_2\}, \tag{4.9}$$

$$\mathbf{K}_G = \sum_{e=1}^{n} \overline{\mathbf{C}}_e^{\mathrm{T}} \mathbf{K}_{Ge} \overline{\mathbf{C}}_e. \tag{4.10}$$

In 3D space, in the geometric stiffness matrix, displacements in three directions (u, v, w) are considered in each node. In the linear stiffness matrix, only one displacement in the direction that is consistent with the bar's axis (u) is defined. Therefore, due to the fact that the number of rows in the Boolean matrices is twice the number of the considered displacements, matrices $\mathbf{C}_e$ and $\overline{\mathbf{C}}_e$ have different dimensions ($m \times 2$ and $m \times 6$ correspondingly).

In a supported structure, vector of nodal displacements $\tilde{\mathbf{q}}$ has a smaller length of $\tilde{n}$. Then, the equations of the linear truss mechanics take a form:

$$\boldsymbol{\Delta} = \tilde{\mathbf{B}}\tilde{\mathbf{q}}, \quad \boldsymbol{\omega} = \tilde{\mathbf{C}}\tilde{\mathbf{q}}, \tag{4.11}$$

$$\mathbf{N} = \mathbf{D}\boldsymbol{\Delta}, \quad \mathbf{F} = \mathbf{S}\boldsymbol{\omega}, \tag{4.12}$$

$$\tilde{\mathbf{B}}^{\mathrm{T}}\mathbf{N} + \tilde{\mathbf{C}}^{\mathrm{T}}\mathbf{F} = \tilde{\mathbf{P}}. \tag{4.13}$$

The system of displacement-based equations for a supported structure is

$$\tilde{\mathbf{K}}\tilde{\mathbf{q}} = \tilde{\mathbf{P}}, \tag{4.14}$$

where:
$\tilde{\mathbf{K}} = \tilde{\mathbf{K}}_{\mathrm{L}} + \tilde{\mathbf{K}}_{\mathrm{G}}$,
$\tilde{\mathbf{K}}_{\mathrm{L}} = \tilde{\mathbf{B}}^{\mathrm{T}}\mathbf{D}\tilde{\mathbf{B}}$,
$\tilde{\mathbf{K}}_{\mathrm{G}} = \tilde{\mathbf{C}}^{\mathrm{T}}\mathbf{S}\tilde{\mathbf{C}}$.

The generalised Maxwell's equation for a supported 3D tensegrity structure has a form

$$m - n = f - s. \tag{4.15}$$

4.2 CONTINUUM MODELS

Continuum models are usually developed to solve some computational problems or reduce complexity of the analysis of structures. In order to build such models, various homogenisation techniques [13] together with an experimental verification [119] are used. Continuum models are applied to analyse various types of structures: bar, plate or shell grids [101,102,132], sandwich plates and shells [137], systems with repeating geometry [20, 38], including micro- [113] and nanostructures [103].

Due to high complexity of tensegrity systems, it is often difficult to estimate their properties using only discrete methods. An interesting approach that makes it possible to determine and understand mechanical characteristics of these unique structures is the application of a continuum model. However, very few works on continuum models of tensegrity systems can be found in the literature. Luo et al. [89] modelled a continuum biological structure with the use of a discrete tensegrity grid. Kebiche et al. [74] proposed a bar model of a tensegrity structure, which took into account the effect of self-stress.

The continuum model of tensegrity systems, which is discussed in this study, was first proposed by Gilewski and Kasprzak [59] and then developed by Al Sabouni-Zawadzka and Gilewski in several works [6, 9, 57]. This model can be successfully used in the recently developed area of tensegrity-inspired metamaterials and lattices [8, 37, 90, 117, 118] as well as for the evaluation of extremal properties of systems [9, 71, 94]. It should be highlighted that the continuum model is not meant to replace the standard approach to the analysis of tensegrity systems—it is just an additional method, which can help to understand and promote this type of structures in engineering applications.

Application of the continuum model is aimed at:

- identification of structural properties corresponding to typical deformation modes (tension, shear) of tensegrity,
- physical interpretation of the determined mechanical characteristics,
- estimation of the influence of self-stress and cable/strut characteristics on deformation of the structure,
- comparison of elastic properties of various tensegrity systems,
- determination of limiting values for the mechanical characteristics,
- identification of extremal mechanical properties of tensegrity structures.

The continuum model is based on the comparison of the strain energy of an unsupported tensegrity structure defined with the use of a discrete model with the strain energy of a solid determined using the symmetric linear 3D elasticity theory (LTE) [62]. The strain energy of a tensegrity truss in a discrete model (DM) is a quadratic form of nodal displacements $\mathbf{q}$:

$$E_{\mathrm{s}}^{\mathrm{DM}} = \frac{1}{2}\mathbf{q}^{\mathrm{T}}\mathbf{K}\mathbf{q}, \qquad (4.16)$$

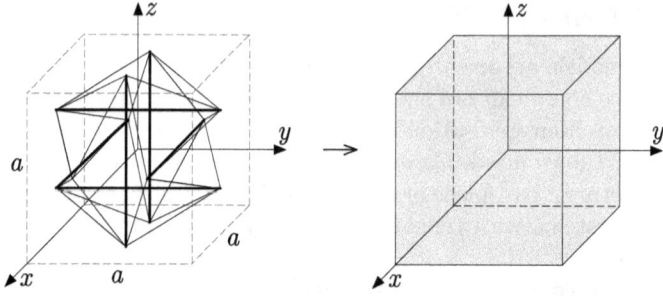

Figure 4.1 Tensegrity module and solid.

where:
$\mathbf{K} = \mathbf{K}_L + \mathbf{K}_G,$
$\mathbf{K}_L$—global linear stiffness matrix,
$\mathbf{K}_G$—global geometric stiffness matrix.

Self-stress of the structure is represented by the geometric stiffness matrix $\mathbf{K}_G$. The stiffness matrices can be determined using FEM [17, 35, 68, 152] or obtained directly (without the finite element approximation), as shown in [85, 110]. However, the FEM formalism is recommended for 3D models [110] as it is simpler to algorithmise.

The strain energy of a solid according to LTE can be expressed as

$$E_s^{\text{LTE}} = \frac{1}{2} \int_V \varepsilon^T \mathbf{E} \varepsilon \, dV, \tag{4.17}$$

where:
ε—strain vector,
$\mathbf{E}$—elasticity matrix.

In order to analyse mechanical properties of tensegrity, the strain energy of an unsupported tensegrity module inscribed into a cube of edge length a is compared with the strain energy of this cube (Figure 4.1). It is assumed that the strain energy of the cube is constant in its whole volume.

With the above assumptions, the strain energy of the cube of edge length a according to LTE is

$$E_s^{\text{LTE}} = \frac{1}{2} \int_V \varepsilon^T \mathbf{E} \varepsilon \, dV = \frac{1}{2} \varepsilon^T \mathbf{E} \varepsilon a^3. \tag{4.18}$$

To compare the energies and build the equivalent elasticity matrix $\mathbf{E}$, the nodal displacements should be expressed by the average mid-values of displacements and their derivatives, using the Taylor series expansion:

$$
\begin{aligned}
f(x + \Delta x, y + \Delta y, z + \Delta z) &= f(x, y, z) + \\
&+ \frac{\partial f}{\partial x}(x, y, z)\Delta x + \frac{\partial f}{\partial y}(x, y, z)\Delta y + \frac{\partial f}{\partial z}(x, y, z)\Delta z + \\
&+ \frac{1}{2}\frac{\partial^2 f}{\partial x^2}(x, y, z)(\Delta x)^2 + \frac{1}{2}\frac{\partial^2 f}{\partial y^2}(x, y, z)(\Delta y)^2 + \frac{1}{2}\frac{\partial^2 f}{\partial z^2}(x, y, z)(\Delta z)^2 + \\
&+ \frac{\partial^2 f}{\partial x \partial y}(x, y, z)\Delta x \Delta y + \frac{\partial^2 f}{\partial x \partial z}(x, y, z)\Delta x \Delta z + \frac{\partial^2 f}{\partial y \partial z}(x, y, z)\Delta y \Delta z + \dots \quad.
\end{aligned} \tag{4.19}
$$

Further analysis is limited to low order terms, what in the case of the first order approximation leads to the first gradient theory within the linear theory of elasticity [62]. Assuming that the nodal coordinates of tensegrity are expressed by the parameter a, the coordinates of node i can be written as $\{\alpha_{xi}a, \alpha_{yi}a, \alpha_{zi}a\}$. Then, the parameters of node i can be described using the average mid-values with the corresponding increments: $\Delta x = \alpha_{xi}a, \Delta y = \alpha_{yi}a, \Delta z = \alpha_{zi}a$.

The described approach leads to the equivalent symmetric elasticity matrix:

$$
\mathbf{H} = \begin{bmatrix}
e_{11} & e_{12} & e_{13} & e_{14} & e_{15} & e_{16} \\
 & e_{22} & e_{23} & e_{24} & e_{25} & e_{26} \\
 & & e_{33} & e_{34} & e_{35} & e_{36} \\
 & & & e_{44} & e_{45} & e_{46} \\
 & & & & e_{55} & e_{56} \\
\text{sym.} & & & & & e_{66}
\end{bmatrix} . \tag{4.20}
$$

Altogether, there are 36 coefficients, including 21 independent ones. The elasticity matrix can take different, particular forms, depending on the type of symmetry. It can be proved that there are exactly eight types of symmetry in the linear theory of elasticity [32]. According to this theory, each material is either isotropic or anisotropic, and among the anisotropic ones the following are distinguished: triclinic (fully anisotropic), monoclinic, trigonal, orthotropic, tetragonal, cubic and transversely isotropic. All these symmetry types have a different number of independent coefficients of matrix $\mathbf{E}$ and a different number of symmetry planes [32, 62].

The discussed continuum model is non-linear in the sense of equilibrium equations considered in actual configuration. Validation of the model for structures with self-stress was presented in the annex to [9].

One of the reasons for using continuum model in the analysis of tensegrity systems is the possibility to identify their mechanical characteristics, which can be expressed by the following technical coefficients:

- Young's moduli (E)—relations between normal stress and strain,
- shear moduli (G)—relations between shear stress and strain,

- Poisson's ratios (ν)—relations between normal strain in perpendicular directions,
- coefficients (μ)—relations between shear strain in perpendicular directions,
- coefficients (λ)—relations between normal strain in three directions and shear strain in one direction,
- coefficients (κ)—relations between shear strain in three directions and normal strain in one direction.

These mechanical characteristics can be calculated from the inverse elasticity matrix $\mathbf{H} = \mathbf{E}^{-1}$, using the constitutive relations from the linear theory of elasticity:

$$\sigma = \mathbf{E}\varepsilon,$$
$$\varepsilon = \mathbf{E}^{-1}\sigma = \mathbf{H}\sigma. \tag{4.21}$$

Analysis of particular stress states makes it possible to determine relations between the values of technical coefficients and the coefficients of matrix $\mathbf{H}$:

$$\mathbf{E} = \begin{bmatrix} \dfrac{1}{E_1} & -\dfrac{\nu_{21}}{E_2} & -\dfrac{\nu_{31}}{E_3} & \dfrac{\lambda_{11}}{G_1} & \dfrac{\lambda_{21}}{G_2} & \dfrac{\lambda_{31}}{G_3} \\[2mm] -\dfrac{\nu_{12}}{E_1} & \dfrac{1}{E_2} & -\dfrac{\nu_{32}}{E_3} & \dfrac{\lambda_{12}}{G_1} & \dfrac{\lambda_{22}}{G_2} & \dfrac{\lambda_{32}}{G_3} \\[2mm] -\dfrac{\nu_{13}}{E_1} & -\dfrac{\nu_{23}}{E_2} & \dfrac{1}{E_3} & \dfrac{\lambda_{13}}{G_1} & \dfrac{\lambda_{23}}{G_2} & \dfrac{\lambda_{33}}{G_3} \\[2mm] \dfrac{\kappa_{11}}{E_1} & \dfrac{\kappa_{21}}{E_2} & \dfrac{\kappa_{31}}{E_3} & \dfrac{1}{G_1} & \dfrac{\mu_{21}}{G_2} & \dfrac{\mu_{31}}{G_3} \\[2mm] \dfrac{\kappa_{12}}{E_1} & \dfrac{\kappa_{22}}{E_2} & \dfrac{\kappa_{32}}{E_3} & \dfrac{\mu_{12}}{G_1} & \dfrac{1}{G_2} & \dfrac{\mu_{32}}{G_3} \\[2mm] \dfrac{\kappa_{13}}{E_1} & \dfrac{\kappa_{23}}{E_2} & \dfrac{\kappa_{33}}{E_3} & \dfrac{\mu_{13}}{G_1} & \dfrac{\mu_{23}}{G_2} & \dfrac{1}{G_3} \end{bmatrix}. \tag{4.22}$$

Due to the fact that both matrices $\mathbf{E}$ and $\mathbf{H}$ have to be positive definite, the following conditions for technical coefficients can be formulated [57]:

$$
\begin{aligned}
& E_1 > 0, \quad E_2 > 0, \quad E_3 > 0, \\
& G_1 > 0, \quad G_2 > 0, \quad G_3 > 0, \\
& \nu_{12}\nu_{21} < 1, \quad \nu_{13}\nu_{31} < 1, \quad \nu_{23}\nu_{32} < 1, \\
& \nu_{12}\nu_{21} + \nu_{13}\nu_{31} + \nu_{23}\nu_{32} + \nu_{12}\nu_{31}\nu_{23} + \nu_{21}\nu_{13}\nu_{32} < 1, \\
& \mu_{12}\mu_{21} < 1, \quad \mu_{13}\mu_{31} < 1, \quad \mu_{23}\mu_{32} < 1, \\
& \mu_{12}\mu_{21} + \mu_{13}\mu_{31} + \mu_{23}\mu_{32} - \mu_{12}\mu_{31}\mu_{23} - \mu_{21}\mu_{13}\mu_{32} < 1.
\end{aligned}
\tag{4.23}
$$

Additionally, principal minors 4×4, 5×5 and 6×6 of matrix $\mathbf{H}$ have to be positive, which leads to global conditions for technical coefficients. These expressions are relatively large and thus, will not be presented in this study. However, they are taken into account in the analyses presented in Section 5.3.2.

The method presented above can be applied to different types of structures (one-, two- and three-dimensional tasks) with any degree of freedom. It can be used to determine properties of simple tensegrity modules as well as more complex multi-module structures, and to evaluate extremal mechanical properties of tensegrity systems, which are thoroughly discussed in Chapter 5.

4.3 SCALE EFFECTS IN THE CONTINUUM MODEL OF TENSEGRITY LATTICES

The continuum model described in Section 4.2 allows us to determine equivalent mechanical properties of tensegrity systems in the range of material symmetries, which are acceptable in the linear theory of elasticity [32, 57]. However, as it was proved by Gilewski and Al Sabouni-Zawadzka in [58], the proposed model can be generalised to models that take into account higher gradients of displacements.

The theories of higher gradients started to appear in the literature in the mid-1960s, first described by Mindlin in [95, 96] and afterwards developed by many researchers. Eringen [42] presented a methodical description of microcontinuum field theories in comparison with other possible formulations. Polyzos and Fotiadis [112] discussed gradient formulations in relation to simple lattice models. Askes and Aifantis [14] published an overview of the formulation of gradient elasticity in statics and dynamics and discussed length scale identification procedures. Forest and Sab [49] focussed on the stress gradient continuum theory.

A continuum model in the field of the higher gradients theory makes it possible to identify the effect of scale, to determine the applicability range of the first approximation theory and to identify problems that should be taken into consideration when dealing with tensegrity structures on a non-micro scale.

Tensegrity systems can be used in various applications, among others as metamaterials with tensegrity microstructure (see Section 2.1) or as modular lattices constructed from regular tensegrity modules (see Section 5.3.2). Both types of systems are created similarly, by connecting multiple tensegrity modules (unit cells) to form bigger structures (Figure 4.2). Depending on the dimensions of the unit cell and the ratio between the size of the unit cell and the whole system, the consideration of higher gradients of displacements will lead to different results. The first approximation model is appropriate when the unit cell is very small and the scale effects should be considered in structures, where the unit cell is relatively large compared to the dimensions of the entire lattice.

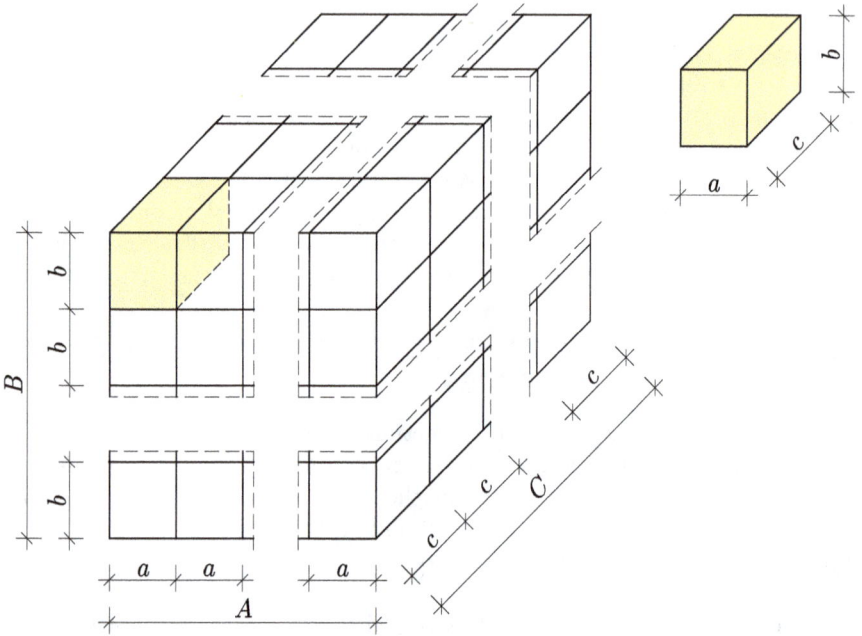

Figure 4.2 Modular tensegrity lattice.

The use of the Taylor series expansion (Eq. 4.19) for all nodal displacements of the structure contained in the unit cell of volume V, allows us to express the density of the strain energy (Eq. 4.16) in the form

$$\tilde{E}_s^{DM} = \frac{1}{V}[E_{s11}(\Delta u) + E_{s12}(\Delta u, \Delta \Delta u) + E_{s22}(\Delta \Delta u) + \dots]. \qquad (4.24)$$

The term "11" depends only on the first displacement gradient, the term "22" only on the second displacement gradient, and the term "12" is a mixed term within the theory of elasticity. The terms of the Eq. 4.24 can be represented as quadratic or bilinear forms:

$$E_{s11}(\Delta u) = \frac{1}{2}(\Delta u)^T E_{11}(\Delta u),$$

$$E_{s12}(\Delta u, \Delta \Delta u) = (\Delta u)^T E_{12}(\Delta \Delta u), \qquad (4.25)$$

$$E_{s22}(\Delta \Delta u) = \frac{1}{2}(\Delta \Delta u)^T E_{22}(\Delta \Delta u).$$

The energy densities of the higher order gradient theory can be determined by analogy. Matrix E_{11} corresponds to the first approximation within the first gradient theory and is insensitive to scale effects. It can be successfully applied to pre-evaluate mechanical properties of tensegrity structures (as shown in [8,57]) and to identify extremal properties of tensegrity modules and lattices [5,9]. The first approximation model provides good results when the unit cell is small. Matrices $\mathbf{E}_{12}$ and $\mathbf{E}_{22}$, on the other hand, correspond to the second gradient theory and are sensitive to scale effects. They can be used to assess behaviour of structures, in which the unit cell is relatively big in comparison with the size of the whole lattice. Matrices $\mathbf{E}_{12}$ and $\mathbf{E}_{22}$ describe the influence of strain variability within the cell (understood as the first gradient of displacements) on mechanical properties of the lattice. Examples of matrices $\mathbf{E}_{11}$, $\mathbf{E}_{12}$, and $\mathbf{E}_{22}$ determined for 2D and 3D tensegrity modules are presented in [58].

5 Extremal Mechanical Properties of Tensegrity Systems

5.1 IDENTIFICATION OF EXTREMAL MECHANICAL PROPERTIES

Extremal properties of tensegrity systems can be investigated in two ways: either by using a discrete approach (see Section 4.1) or by applying a continuum model (see Section 4.2). While discrete models are suitable for determination of extremal mechanical properties of metastructures, that is tensegrity lattices in a structural scale, the continuum approach allows us to investigate extremal behaviour of metamaterials.

Let us introduce two parameters: k and σ, that will be used for the adjustment of mechanical properties of tensegrities while searching for extremal properties of the considered systems:

$$k = \frac{(EA)_{\text{cable}}}{(EA)_{\text{strut}}}, \quad (EA)_{\text{strut}} = EA, \quad \sigma = \frac{S_0}{EA}, \quad (5.1)$$

where:
k—cable-to-strut stiffness ratio,
σ—level of self-stress,
E—Young's modulus,
A—cross-sectional area of the member,
S_0—multiplier of self-stress forces.

In the discrete approach, the procedure for determining extremal mechanical properties of tensegrity metastructures is based on spectral analysis of the stiffness matrix $\mathbf{K} = \mathbf{K}_{\text{L}} + \mathbf{K}_{\text{G}}$:

$$(\mathbf{K}_{\text{L}} + \mathbf{K}_{\text{G}} - \chi \mathbf{I}) \, \mathbf{q} = 0. \quad (5.2)$$

Both matrices $\mathbf{K}_{\text{L}}$ and $\mathbf{K}_{\text{G}}$ depend on parameters k and σ. All eigenvalues χ which are equal to zero (apart from the zero values corresponding to rigid movements) indicate soft modes of deformation, that is, deformation states in which the system is extremely compliant. Such deformation modes are expressed by eigenvectors $\mathbf{q}_{\text{soft}}$, which generate the strain energy equal to zero. The following equation is satisfied:

$$\frac{1}{2} \mathbf{q}_{\text{soft}}^{\text{T}} (\mathbf{K}_{\text{L}} + \mathbf{K}_{\text{G}}) \, \mathbf{q}_{\text{soft}} = 0. \quad (5.3)$$

DOI: 10.1201/9781003343202-5

Such an approach allows us to find pairs of parameters k_{extr} and σ_{extr}, for which the soft modes of deformation occur. However, the use of a discrete model has certain disadvantages. Dimensions of stiffness matrices, even in simple tasks, are usually too big to be able to perform spectral analysis at a reasonable computational cost. Therefore, some other approaches should be developed for the investigation of extremal behaviour of metastructures. One of possible procedures is proposed further in this book (see Section 5.6).

In a material scale that is in the case of tensegrity-inspired metamaterials, determination of extremal mechanical properties can be performed using a continuum model. Such an approach allows us to spare a lot of computational time and is more elegant from the mathematical point of view. This approach is based on the elasticity matrix $\mathbf{E}$ determined using the continuum model (see Section 4.2). The matrix has dimensions of 3×3 in the case of 2D systems, and 6×6 in the case of 3D systems, and it depends on two parameters: k and σ. What is very important, the dimensions of matrix $\mathbf{E}$ do not depend on the task complexity, they are the same for a single module and a very complex metamaterial. After solving the eigenvalue problem

$$(\mathbf{E} - \lambda \mathbf{I})\,\mathbf{w} = \mathbf{0}, \tag{5.4}$$

it is possible to determine the values of parameters k_{extr} and σ_{extr}, which ensure occurrence of soft modes of deformation—all eigenvalues λ which are equal to zero (apart from the zero values corresponding to rigid movements) indicate soft modes of deformation. Corresponding eigenvectors $\mathbf{w}$, on the other hand, describe deformation states—starting from soft modes in which the system is very compliant, up to stiff modes where the structure is extremely stiff.

This chapter is focussed on a thorough analysis of extremal mechanical properties of various tensegrity systems. For each system a continuum model is built, which is used for finding the elasticity matrix $\mathbf{E}$ of the given structure. Afterwards, a spectral analysis of matrix $\mathbf{E}$ is performed, which allows us to find eigenvalues and eigenvectors describing extremal mechanical properties of the system, such as soft, medium and stiff modes of deformation.

The following systems are considered:

- systems in 3D space:
  - 3D tensegrity modules,
  - 3D modular tensegrity lattices,
- systems in 2D space:
  - 2D tensegrity modules,
  - 2D modular tensegrity lattices.

Among 3D tensegrity modules, five basic structures are considered: three-strut simplex, four-strut simplex, expanded octahedron, truncated tetrahedron and X-module, each in five geometrical variants. Then, the modules are used to build modular tensegrity lattices, out of which three four-strut simplex lattices are described and analysed.

In the second part of this chapter, the main focus is put on the analysis of extremal mechanical properties of five 2D tensegrity modules: two hexagonal and three octagonal modules. The analysis of 2D tensegrity lattices is limited to the discussion on possible configurations of the systems, as the properties of these lattices are the same as the properties of their basic cells.

The chapter is organised in this way—3D systems are discussed before 2D systems—because 3D modules and lattices are more important from the practical point of view, especially when bigger scales are to be considered. Therefore, 3D tensegrity metamaterials and metastructures are the main focus of this study. However, in structural mechanics, 2D computational models are often used to simulate complex engineering phenomena in a simpler, more understandable way—the solutions obtained in this way are accurate enough for engineering purposes (e.g. design of structures). In the case of extremal metamaterials and metastructures, analysis of 2D systems adds value to the understanding of extremal mechanical properties, as it leads to closed-form solutions for eigenvalues that describe these properties.

At the end, one selected 2D module (hexagon 1) and one 3D module (four-strut simplex) are used to study scale effects that occur while moving from a material scale (metamaterial) to a structural one (metastructure). Results from the continuum model are applied as an input in the discrete analysis of the modules, and it is proved that extremal mechanical properties of the considered systems occur not only in the metamaterial scale, but also in the case of metastructures.

5.2 TENSEGRITY MODULES IN 3D SPACE

In this section, a study on extremal mechanical properties of five basic tensegrity modules is presented: three-strut simplex, four-strut simplex, expanded octahedron, truncated tetrahedron, and X-module (Figure 5.1).

Each module is analysed in five variants of geometrical proportions:

- regular—the module is inscribed into a cube $a \times a \times a$,
- high—the module is inscribed into a cuboid $\frac{1}{2}a \times \frac{1}{2}a \times a$,
- very high—the module is inscribed into a cuboid $\frac{1}{3}a \times \frac{1}{3}a \times a$,

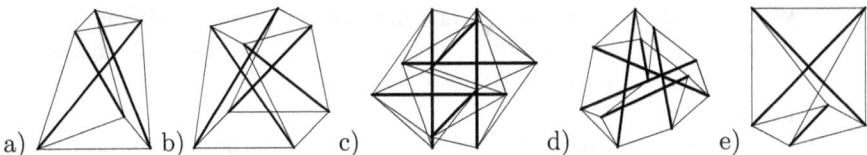

Figure 5.1 Basic tensegrity modules: a) three-strut simplex; b) four-strut simplex; c) expanded octahedron; d) truncated tetrahedron; e) X-module.

- low—the module is inscribed into a cuboid $a \times a \times \frac{1}{2}a$,
- very low—the module is inscribed into a cuboid $a \times a \times \frac{1}{3}a$.

The analyses are based on the control of two parameters k and σ (parameters introduced in Section 5.1), which can be adjusted while searching for extremal properties of the considered structures. In the analyses presented in Sections 5.2.1–5.2.5, parameter k is assumed constant and equal to 0.1 for all presented cases, which makes it possible to compare the considered modules. In Section 5.2.6 an influence of parameter k on extremal properties of selected tensegrity modules is analysed and the issue of applied material properties is discussed.

In order to analyse extremal mechanical properties of the structures, the following features are described for regular modules:

- equivalent elasticity matrix,
- extremal properties:
  - the line on plane (k, σ), indicating pairs of parameters for which the smallest eigenvalue of the elasticity matrix is close to zero—this line indicates a possible occurrence of the soft mode of deformation (an arrow shows the half-plane for which matrix $\mathbf{E}$ is positive definite),
  - distribution of eigenvalues for $k_{\text{extr}} = 0.1$ and the corresponding value of σ_{extr}, scaled so that the volumes of all modules are identical and equal to the volume of the three-strut simplex module, assuming the same parent material of cables and struts,
  - eigenvectors corresponding to individual eigenvalues of matrix $\mathbf{E}$—the eigenvectors corresponding to soft and stiff deformation modes are additionally presented in the drawings.

For other variants—high, very high, low and very low modules—the presented results are limited to the distribution of eigenvalues obtained for $k_{\text{extr}} = 0.1$ and the corresponding value of σ_{extr}. In these cases, however, the values are not scaled—it is assumed that for each module its four variants are constructed from the same struts and cables as the regular one.

Determination of eigenvalues makes it possible to identify soft, medium and stiff modes of deformation. In the presented analyses it is assumed that:

- the eigenvalues, which are smaller than 1% of the maximum eigenvalue obtained for the given module, indicate purely soft modes of deformation—such structures are referred to as unimode, bimode or trimode (such an approximation is accurate from the engineering point of view),
- the eigenvalues, which are between 1% and 2% of the maximum eigenvalue obtained for the given module, indicate quasi soft modes of deformation—such structures are referred to as quasi unimode, quasi bimode or quasi trimode.

Table 5.1

Geometry of the regular three-strut simplex.

Node No.	x	y	z	Scheme
1	$0.366 \cdot a$	$-0.500 \cdot a$	$-0.500 \cdot a$	
2	$0.366 \cdot a$	$0.500 \cdot a$	$-0.500 \cdot a$	
3	$-0.500 \cdot a$	0	$-0.500 \cdot a$	
4	$0.366 \cdot a$	$-0.167 \cdot a$	$0.500 \cdot a$	
5	$0.077 \cdot a$	$0.333 \cdot a$	$0.500 \cdot a$	
6	$-0.211 \cdot a$	$-0.167 \cdot a$	$0.500 \cdot a$	

5.2.1   THREE-STRUT SIMPLEX FAMILY

The analysed regular three-strut simplex module (S3) is presented in Figure 5.2. Geometry of the module is described in Table 5.1, which contains nodal coordinates of the three-strut simplex.

The module has one infinitesimal mechanism and one corresponding self-stress state (Figure 5.3)—self-stress is expressed by relative forces in struts and cables with a multiplier S_0.

The elasticity matrix obtained from the continuum model has a form

$$
\mathbf{E}_{S3} = \begin{bmatrix}
3e_{12} & e_{12} & e_{13} & 0 & e_{15} & e_{16} \\
 & 3e_{12} & e_{13} & 0 & -e_{15} & -e_{16} \\
 & & e_{33} & 0 & 0 & 0 \\
 & & & e_{12} & e_{16} & -e_{15} \\
 & & & & e_{13} & 0 \\
\text{sym.} & & & & & e_{13}
\end{bmatrix},
\tag{5.5}
$$

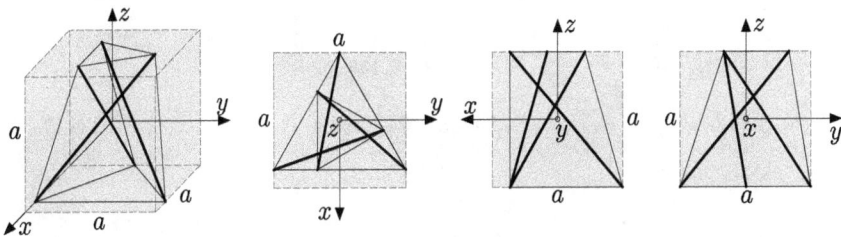

Figure 5.2 Regular three-strut simplex module inscribed into a cube.

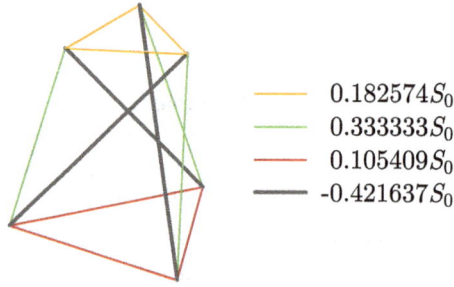

Figure 5.3 Self-stress state in the regular three-strut simplex.

$$e_{12} = \frac{EA}{a^2}(0.0957031 + 0.595459 \cdot k - 0.0400226 \cdot \sigma),$$

$$e_{13} = \frac{EA}{a^2}(0.492188 + 0.142302 \cdot k + 0.16009 \cdot \sigma),$$

$$e_{15} = \frac{EA}{a^2}(0.182677 + 0.0770235 \cdot \sigma), \tag{5.6}$$

$$e_{16} = \frac{EA}{a^2}(-0.117187 - 0.0237171 \cdot k - 0.0415049 \cdot \sigma),$$

$$e_{33} = \frac{2EA}{a^2}(0.632813 + 1.28072 \cdot k - 0.16009 \cdot \sigma).$$

The elasticity matrix indicates that the regular three-strut simplex has anisotropic properties. Spectral analysis of this matrix allows us to find eigenvalues and corresponding eigenvectors, which describe extremal properties of the module. They cannot be determined explicitly as a function of k and σ, but it is possible to identify extremal properties for the given pairs of these parameters. Let us adopt $k_{\text{extr}} = 0.1$, then it is possible to determine the value of $\sigma_{\text{extr}} = 0.5635$, which leads to the following eigenvalues and corresponding eigenvectors:

$$\lambda_1 = 1.87215\frac{EA}{a^2} : \quad \mathbf{w}_{1,S3} = [0.444759 \quad 0.444759 \quad 1 \quad 0 \quad 0 \quad 0]^{\mathrm{T}},$$

$$\lambda_2 = 0.84385\frac{EA}{a^2} : \quad \mathbf{w}_{2,S3} = [0.547031 \quad -0.547031 \quad 0 \quad 0 \quad 1 \quad -0.632300]^{\mathrm{T}},$$

$$\lambda_3 = 0.71862\frac{EA}{a^2} : \quad \mathbf{w}_{3,S3} = [0 \quad 0 \quad 0 \quad -0.540043 \quad 0.632300 \quad 1]^{\mathrm{T}},$$

$$\lambda_4 = 0.01813\frac{EA}{a^2} : \quad \mathbf{w}_{4,S3} = [-1 \quad 1 \quad 0 \quad 0 \quad 0.781582 \quad 0.494195]^{\mathrm{T}},$$

$$\lambda_5 = 0.01064\frac{EA}{a^2} : \quad \mathbf{w}_{5,S3} = [0 \quad 0 \quad 0 \quad 1 \quad 0.243941 \quad 0.385799]^{\mathrm{T}},$$

$$\lambda_6 = 0 : \quad \mathbf{w}_{6,S3} = [1 \quad 1 \quad -0.889519 \quad 0 \quad 0 \quad 0]^{\mathrm{T}}.$$

$$\tag{5.7}$$

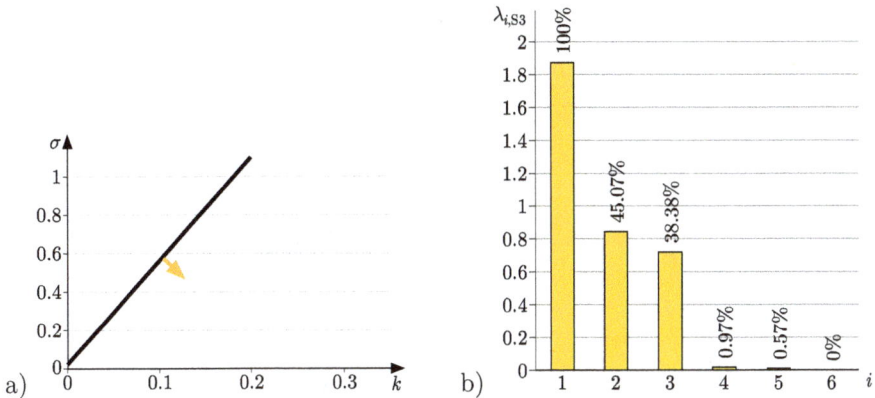

Figure 5.4 Extremal properties of the regular three-strut simplex: a) line of extremal properties $\sigma = 0.023 + 5.4 \cdot k$; b) distribution of eigenvalues for $k_{extr} = 0.1$, $\sigma_{extr} = 0.5635$ (multiplier EA/a^2).

It is worth mentioning that for different values of k_{extr}, different values of σ_{extr} would be obtained. However, although each pair of parameters k_{extr} and σ_{extr} leads to different eigenvalues, the eigenvectors describing extremal modes of deformation remain unchanged. Extremal properties of the module are presented in Figure 5.4.

The module can be identified as trimode, since one eigenvalue is equal to zero and two others are smaller than 1% of the maximum eigenvalue, what—from the engineering point of view—can be approximated to zero. One positive eigenvalue is dominant over the others. Analysis of the graphs (Figure 5.4) leads to the conclusion that the regular three-strut simplex module has one stiff, three soft (one zero eigenvalue and two eigenvalues less than 1% of $\lambda_{max,S3}$) and two medium (eigenvalues around 38% and 45% of $\lambda_{max,S3}$) modes of deformation.

Extremal properties of the module are expressed by stiff (represented by the eigenvector $\mathbf{w}_{1,S3}$) and soft (represented by the eigenvectors $\mathbf{w}_{4,S3}$, $\mathbf{w}_{5,S3}$ and $\mathbf{w}_{6,S3}$) modes of deformation and therefore, medium deformation modes are not further analysed.

The stiff mode (Figure 5.5a), represented by the eigenvector $\mathbf{w}_{1,S3}$, is volumetric with a uniform sign, and the extension in X_3 direction is 125% bigger than in others. The soft mode (Figure 5.5b), represented by the eigenvector $\mathbf{w}_{4,S3}$, is a combination of a shear deformation dominant in X_1–X_3 plane and a volumetric deformation with the contraction in X_1 direction equal to the extension in X_2 direction. The soft mode (Figure 5.5c), represented by the eigenvector $\mathbf{w}_{5,S3}$, is a shear deformation dominant in X_1–X_2 plane. The soft mode (Figure 5.5d), represented by the eigenvector $\mathbf{w}_{6,S3}$, is volumetric with various signs, and the contraction in X_3 direction is 11% lower than the extension in other directions.

$$\mathbf{w}_{1,\mathrm{S3}} = \begin{bmatrix} 0.444759 & 0.444759 & 1 & 0 & 0 & 0 \end{bmatrix}^{\mathrm{T}}$$

$$\mathbf{w}_{4,\mathrm{S3}} = \begin{bmatrix} -1 & 1 & 0 & 0 & 0.781582 & 0.494195 \end{bmatrix}^{\mathrm{T}}$$

$$\mathbf{w}_{5,\mathrm{S3}} = \begin{bmatrix} 0 & 0 & 0 & 1 & 0.243941 & 0.385799 \end{bmatrix}^{\mathrm{T}}$$

$$\mathbf{w}_{6,\mathrm{S3}} = \begin{bmatrix} 1 & 1 & -0.889519 & 0 & 0 & 0 \end{bmatrix}^{\mathrm{T}}$$

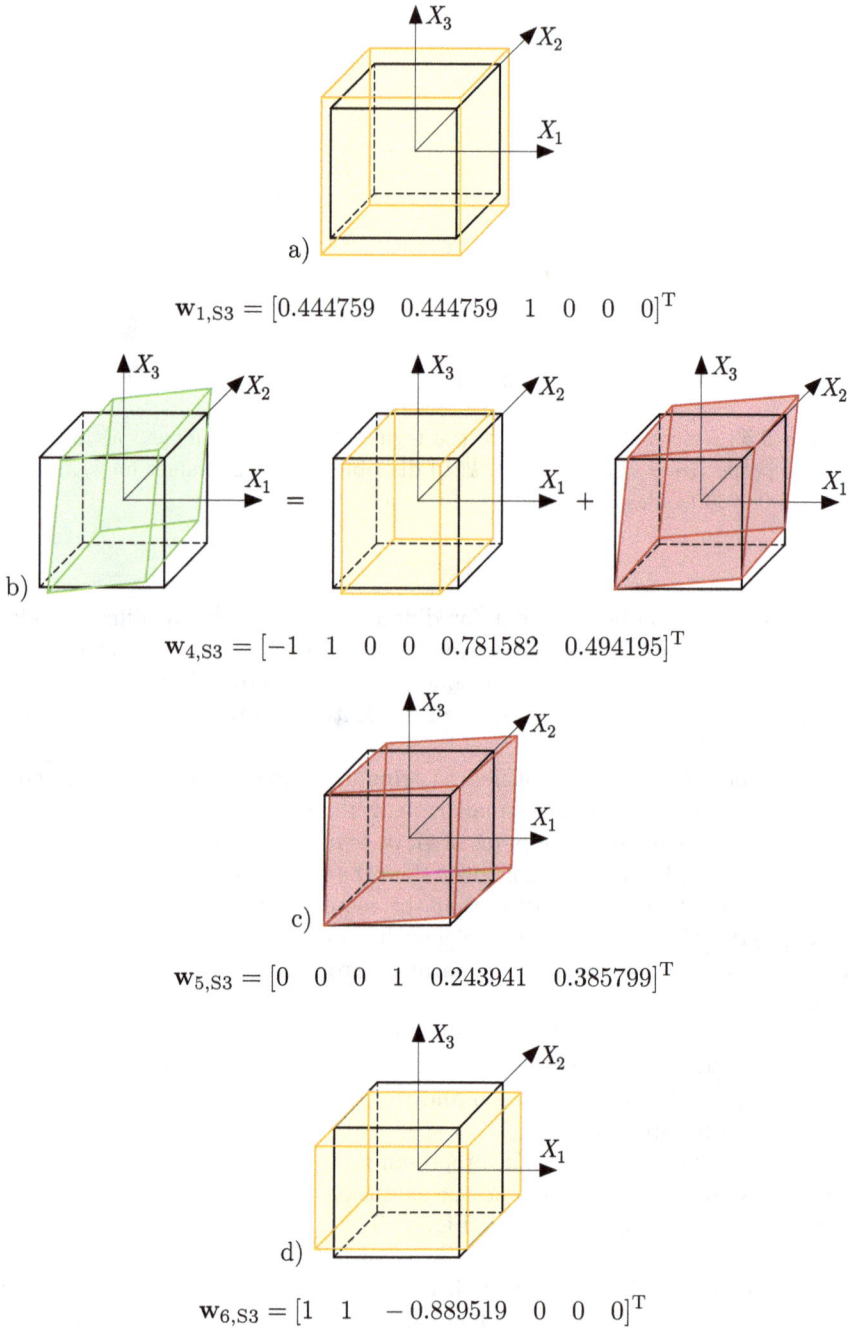

Figure 5.5 Deformation modes of the regular three-strut simplex: a) stiff mode corresponding to λ_1; b) soft mode corresponding to λ_4; c) soft mode corresponding to λ_5; d) soft mode corresponding to λ_6.

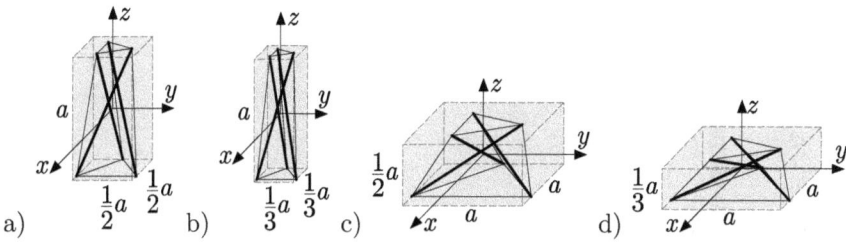

Figure 5.6 Variants of the three-strut simplex module: a) high; b) very high; c) low; d) very low.

Apart from the regular module inscribed into a cube of edge length a, four other variants of the three-strut simplex module were analysed (Figure 5.6), with the proportions described in Section 5.2.

The elasticity matrices obtained for the considered variants of the three-strut simplex have the same form as the elasticity matrix of the regular module $\mathbf{E}_{S3}$ (Eq. 5.5), but with different coefficients e_{ij}. The matrices are not presented here, as the main aim of this section is to discuss extremal properties of the modules and the determination of elasticity matrices is just a step of the analysis leading to the identification of soft, medium and stiff modes of deformation.

Figure 5.7 presents distributions of eigenvalues obtained for five variants of the three-strut simplex. For each eigenvalue a percentage value is given—the eigenvalue corresponding to the stiff mode is assumed as a reference value of 100% and for the following modes a ratio of $\lambda_{i,S3}/\lambda_{\max,S3} \cdot 100\%$ is calculated. The percentage values marked with colours indicate soft modes of deformation (the colours correspond to the module variants). Extremal properties presented in Figure 5.7 were obtained for the parameters $k_{\text{extr}} = 0.1$ and σ_{extr} given in Table 5.2, depending on the analysed variant.

All variants of the three-strut simplex can be identified as trimode, since in each case, three eigenvalues are smaller than 1% of the maximum eigenvalue obtained for the given variant. The variant denoted as very high exhibits the

Table 5.2

Extremal properties of five three-strut simplex variants ($k_{\text{extr}} = 0.1$).

Module	σ_{extr}	Extremal Properties
regular	0.5635	trimode
high	1.0718	trimode
very high	1.6416	trimode
low	0.4459	trimode
very low	0.4712	trimode

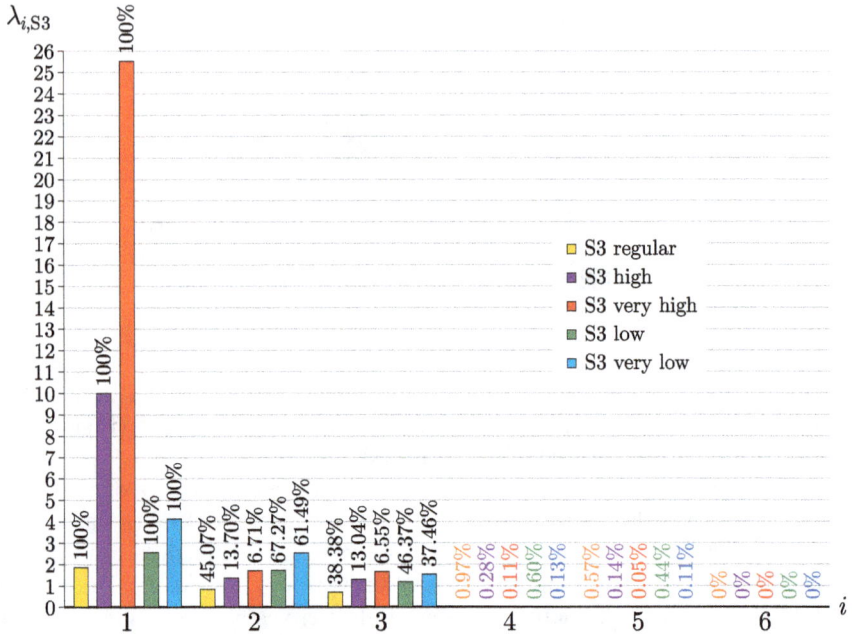

Figure 5.7 Distribution of eigenvalues for five variants of the three-strut simplex module (multiplier EA/a^2).

biggest differences between the eigenvalue corresponding to the stiff mode and the ones indicating medium modes of deformation (eigenvalues around 6.5% of $\lambda_{\mathrm{max,S3}}$).

5.2.2 FOUR-STRUT SIMPLEX FAMILY

The analysed regular four-strut simplex module (S4) is presented in Figure 5.8. Geometry of the module is described in Table 5.3, which contains nodal coordinates of the four-strut simplex.

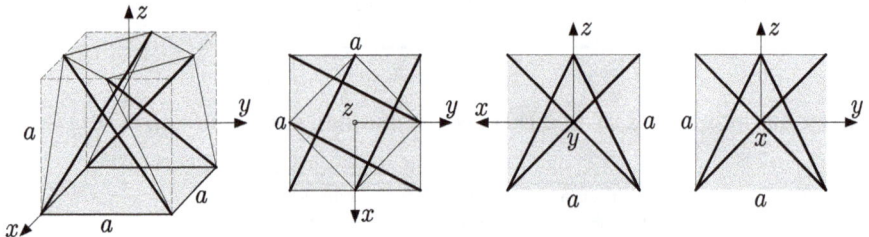

Figure 5.8 Regular four-strut simplex module inscribed into a cube.

Table 5.3

Geometry of the regular four-strut simplex.

Node No.	x	y	z	Scheme
1	$0.5 \cdot a$	$-0.5 \cdot a$	$-0.5 \cdot a$	
2	$0.5 \cdot a$	$0.5 \cdot a$	$-0.5 \cdot a$	
3	$-0.5 \cdot a$	$0.5 \cdot a$	$-0.5 \cdot a$	
4	$-0.5 \cdot a$	$-0.5 \cdot a$	$-0.5 \cdot a$	
5	0	$-0.5 \cdot a$	$0.5 \cdot a$	
6	$0.5 \cdot a$	0	$0.5 \cdot a$	
7	0	$0.5 \cdot a$	$0.5 \cdot a$	
8	$-0.5 \cdot a$	0	$0.5 \cdot a$	

The module has one infinitesimal mechanism and one self-stress state (Figure 5.9)—self-stress is expressed by relative forces in struts and cables with a multiplier S_0.

The elasticity matrix obtained from the continuum model has a form

$$
\mathbf{E}_{S4} =
\begin{bmatrix}
e_{11} & e_{12} & e_{13} & e_{14} & 0 & 0 \\
 & e_{11} & e_{13} & -e_{14} & 0 & 0 \\
 & & e_{33} & 0 & 0 & 0 \\
 & & & e_{12} & 0 & 0 \\
 & & & & e_{13} & 0 \\
\text{sym.} & & & & & e_{13}
\end{bmatrix},
\qquad (5.8)
$$

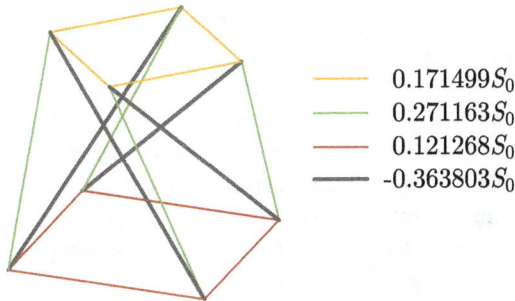

——	$0.171499 S_0$
——	$0.271163 S_0$
——	$0.121268 S_0$
▬▬	$-0.363803 S_0$

Figure 5.9 Self-stress state in the regular four-strut simplex.

$$e_{11} = \frac{2EA}{a^2}(0.314815 + 1.39827 \cdot k - 0.0794978 \cdot \sigma),$$

$$e_{12} = \frac{EA}{a^2}(0.296296 + 0.707107 \cdot k - 0.0134742 \cdot \sigma),$$

$$e_{13} = \frac{EA}{a^2}(0.740741 + 0.357771 \cdot k + 0.17247 \cdot \sigma), \qquad (5.9)$$

$$e_{14} = \frac{EA}{a^2}(-0.222222 - 0.0808452 \cdot \sigma),$$

$$e_{33} = \frac{2EA}{a^2}(0.592593 + 1.43108 \cdot k - 0.17247 \cdot \sigma).$$

Similarly to the previous module, the elasticity matrix indicates that the regular four-strut simplex has anisotropic properties. Spectral analysis of this matrix allows us to find eigenvalues and corresponding eigenvectors, which describe extremal properties of the module. The identification of extremal mechanical behaviour was performed in the same way as described in Section 5.2.1. The following eigenvalues and corresponding eigenvectors were obtained for $k_{\text{extr}} = 0.1$ and $\sigma_{\text{extr}} = 0.5464$:

$$\lambda_1 = 2.46496 \frac{EA}{a^2} : \quad \mathbf{w}_{1,S4} = [0.678717 \quad 0.678717 \quad 1 \quad 0 \quad 0 \quad 0]^T,$$

$$\lambda_2 = 0.87069 \frac{EA}{a^2} : \quad \mathbf{w}_{2,S4} = [0 \quad 0 \quad 0 \quad 0 \quad 1 \quad 0]^T,$$

$$\lambda_3 = 0.87069 \frac{EA}{a^2} : \quad \mathbf{w}_{3,S4} = [0 \quad 0 \quad 0 \quad 0 \quad 0 \quad 1]^T,$$

$$\lambda_4 = 0.79145 \frac{EA}{a^2} : \quad \mathbf{w}_{4,S4} = [-0.810541 \quad 0.810541 \quad 0 \quad 1 \quad 0 \quad 0]^T,$$

$$\lambda_5 = 0.03103 \frac{EA}{a^2} : \quad \mathbf{w}_{5,S4} = [0.616872 \quad -0.616872 \quad 0 \quad 1 \quad 0 \quad 0]^T,$$

$$\lambda_6 = 0 : \quad \mathbf{w}_{6,S4} = [-0.736684 \quad -0.736684 \quad 1 \quad 0 \quad 0 \quad 0]^T. \qquad (5.10)$$

Extremal properties of the module are presented in Figure 5.10.

The module can be identified as quasi bimode, since one eigenvalue is equal to zero and the second is much smaller than the other four. One positive eigenvalue is dominant over the others. Analysis of the graphs (Figure 5.10) leads to the conclusion that the regular four-strut simplex module has one stiff, two soft (one purely soft with the corresponding zero eigenvalue, and one quasi soft with the eigenvalue less than 1.3% of $\lambda_{\text{max,S4}}$) and three medium (eigenvalues around 32% and 35% of $\lambda_{\text{max,S4}}$) modes of deformation.

The stiff mode (Figure 5.11a), represented by the eigenvector $\mathbf{w}_{1,S4}$, is volumetric with a uniform sign, and the extension in X_3 direction is 47% bigger than in others. The quasi soft mode (Figure 5.11b), represented by the

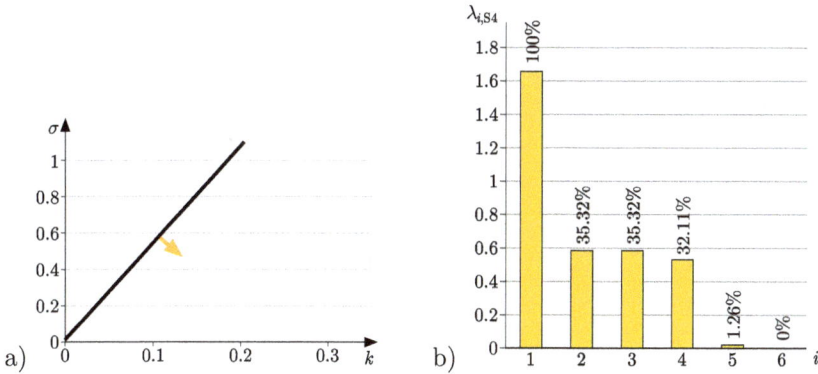

Figure 5.10 Extremal properties of the regular four-strut simplex: a) line of extremal properties $\sigma = 0.012 + 5.34 \cdot k$; a) distribution of eigenvalues for $k_{\text{extr}} = 0.1$, $\sigma_{\text{extr}} = 0.5464$ (multiplier EA/a^2).

eigenvector $\mathbf{w}_{5,\text{S4}}$, is a combination of a shear deformation in X_1–X_2 plane and a volumetric deformation with the extension in X_1 direction equal to the contraction in X_2 direction. The soft mode (Figure 5.11c), represented by the eigenvector $\mathbf{w}_{6,\text{S4}}$, is volumetric with various signs, and the extension in X_3 direction exceeds by 35% the contraction in other directions.

Apart from the regular module inscribed into a cube of edge length a, four other variants of the four-strut simplex module were analysed (Figure 5.12), with the proportions described in Section 5.2.

The elasticity matrices obtained for the considered variants of the four-strut simplex have the same form as the elasticity matrix of the regular module $\mathbf{E}_{\text{S4}}$ (Eq. 5.8), but with different coefficients e_{ij}.

Figure 5.13 presents distributions of eigenvalues obtained for five variants of the four-strut simplex. For each eigenvalue a percentage value is given—the eigenvalue corresponding to the stiff mode is assumed as a reference value of 100% and for the following modes a ratio of $\lambda_{i,\text{S4}}/\lambda_{\text{max},\text{S4}} \cdot 100\%$ is calculated. The percentage values marked with colours indicate soft modes of deformation (the colours correspond to the module variants). Extremal properties presented in Figure 5.13 were obtained for the parameters $k_{\text{extr}} = 0.1$ and σ_{extr} given in Table 5.4, depending on the analysed variant.

While the regular four-strut simplex is quasi bimode, three other variants of this module (high, low, and very low) can be identified as bimode, since in each case, two eigenvalues are smaller than 1% of the maximum eigenvalue obtained for the given variant, and one variant (very high) can be identified as quasi trimode, since two eigenvalues are smaller than 1% and one is smaller than 2% of the maximum eigenvalue. The variant denoted as very high exhibits the biggest differences between the eigenvalue corresponding to the stiff mode and the ones indicating medium modes of deformation (eigenvalues around 9% of $\lambda_{\text{max},\text{S4}}$).

$$\mathbf{w}_{1,S4} = [0.678717 \quad 0.678717 \quad 1 \quad 0 \quad 0 \quad 0]^{\mathrm{T}}$$

$$\mathbf{w}_{5,S4} = [0.616872 \quad -0.616872 \quad 0 \quad 1 \quad 0 \quad 0]^{\mathrm{T}}$$

$$\mathbf{w}_{6,S4} = [-0.736684 \quad -0.736684 \quad 1 \quad 0 \quad 0 \quad 0]^{\mathrm{T}}$$

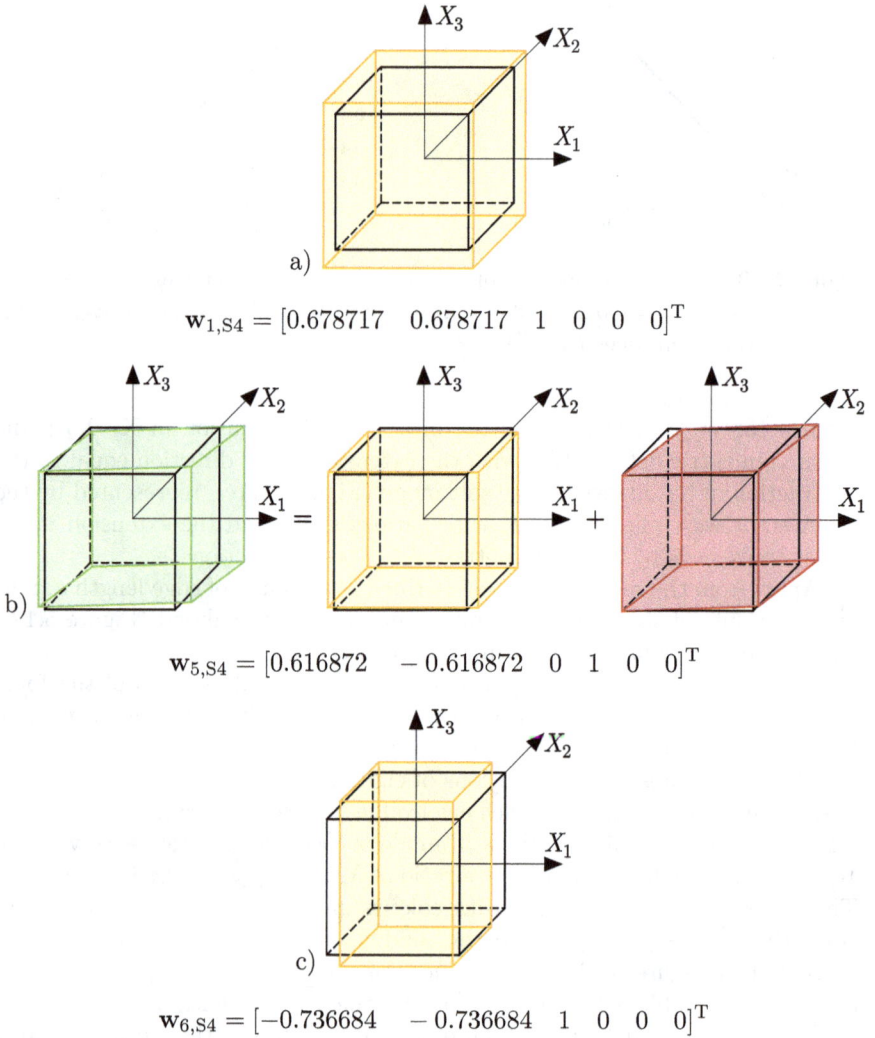

Figure 5.11 Deformation modes of the regular four-strut simplex: a) stiff mode corresponding to λ_1; b) quasi soft mode corresponding to λ_5; c) soft mode corresponding to λ_6.

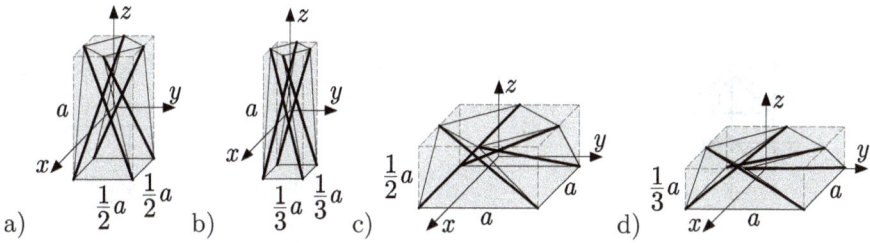

Figure 5.12 Variants of the four-strut simplex module: a) high; b) very high; c) low; d) very low.

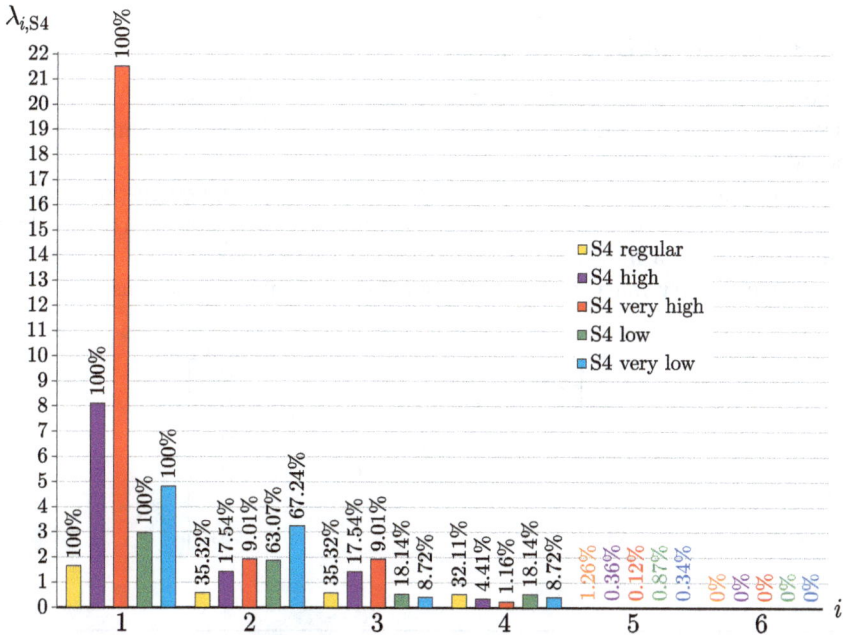

Figure 5.13 Distribution of eigenvalues for five variants of the four-strut simplex module (multiplier EA/a^2).

Table 5.4

Extremal properties of five four-strut simplex variants ($k_{extr} = 0.1$).

Module	σ_{extr}	Extremal Properties
regular	0.5464	quasi bimode
high	0.9150	bimode
very high	1.3734	quasi trimode
low	0.5000	bimode
very low	0.5268	bimode

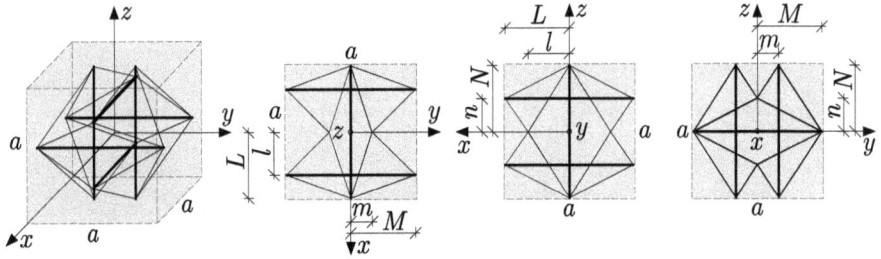

Figure 5.14 Regular expanded octahedron module inscribed into a cube, with adopted parameters: $l/L = 0.65, m/M = 0.30, n/N = 0.56$.

5.2.3 EXPANDED OCTAHEDRON FAMILY

The analysed regular expanded octahedron module (O) is presented in Figure 5.14. Geometry of the module is described in Table 5.5, which contains nodal coordinates of the expanded octahedron.

The module has one infinitesimal mechanism and one corresponding self-stress state (Figure 5.15)—self-stress is expressed by relative forces in struts and cables with a multiplier S_0.

The elasticity matrix obtained from the continuum model has a form

$$
\mathbf{E_O} = \begin{bmatrix}
e_{11} & e_{12} & e_{13} & 0 & 0 & 0 \\
 & e_{22} & e_{23} & 0 & 0 & 0 \\
 & & e_{33} & 0 & 0 & 0 \\
 & & & e_{12} & 0 & 0 \\
 & & & & e_{13} & 0 \\
\text{sym.} & & & & & e_{23}
\end{bmatrix}, \tag{5.11}
$$

Table 5.5
Geometry of the regular expanded octahedron.

Node No.	x	y	z	Scheme
1	$-0.325 \cdot a$	$-0.500 \cdot a$	0	
2	$0.325 \cdot a$	$-0.500 \cdot a$	0	
3	$-0.325 \cdot a$	$0.500 \cdot a$	0	
4	$0.325 \cdot a$	$0.500 \cdot a$	0	
5	0	$-0.150 \cdot a$	$-0.500 \cdot a$	
6	0	$0.150 \cdot a$	$-0.500 \cdot a$	
7	0	$-0.150 \cdot a$	$0.500 \cdot a$	
8	0	$0.150 \cdot a$	$0.500 \cdot a$	
9	$-0.500 \cdot a$	0	$-0.278 \cdot a$	
10	$-0.500 \cdot a$	0	$0.278 \cdot a$	
11	$0.500 \cdot a$	0	$-0.278 \cdot a$	
12	$0.500 \cdot a$	0	$0.278 \cdot a$	

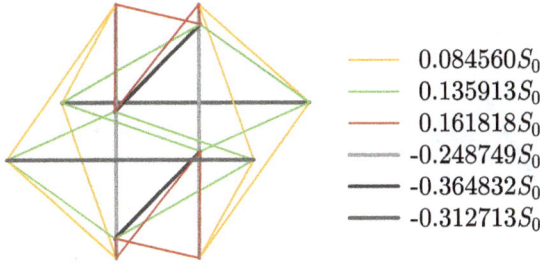

Figure 5.15 Self-stress state in the regular expanded octahedron.

$$e_{11} = \frac{2EA}{a^2}(1 + 1.52325 \cdot k + 0.129225 \cdot \sigma),$$

$$e_{12} = \frac{EA}{a^2}(0.845615 \cdot k - 0.105243 \cdot \sigma),$$

$$e_{13} = \frac{EA}{a^2}(1.26604 \cdot k - 0.153207 \cdot \sigma),$$

$$e_{22} = \frac{2EA}{a^2}(1 + 1.35912 \cdot k + 0.137028 \cdot \sigma),$$

$$e_{23} = \frac{EA}{a^2}(1.51283 \cdot k - 0.168813 \cdot \sigma),$$

$$e_{33} = \frac{2EA}{a^2}(1 + 0.921194 \cdot k + 0.16101 \cdot \sigma).$$

$$(5.12)$$

In this case, the elasticity matrix indicates that the regular expanded octahedron has orthotropic properties. Spectral analysis of this matrix allows us to find eigenvalues and corresponding eigenvectors, which describe extremal properties of the module. The identification of extremal mechanical behaviour was performed in the same way as described in Section 5.2.1. The following eigenvalues and corresponding eigenvectors were obtained for $k_{\text{extr}} = 0.1$ and $\sigma_{\text{extr}} = 0.8034$:

$$\lambda_1 = 2.51250\frac{EA}{a^2} : \quad \mathbf{w}_{1,O} = [1 \quad 0.0506881 \quad 0.0628794 \quad 0 \quad 0 \quad 0]^{\text{T}},$$

$$\lambda_2 = 2.49651\frac{EA}{a^2} : \quad \mathbf{w}_{2,O} = [-0.068845 \quad 1 \quad 0.288761 \quad 0 \quad 0 \quad 0]^{\text{T}},$$

$$\lambda_3 = 2.43822\frac{EA}{a^2} : \quad \mathbf{w}_{3,O} = [-0.048075 \quad -0.292070 \quad 1 \quad 0 \quad 0 \quad 0]^{\text{T}},$$

$$\lambda_4 = 0.01566\frac{EA}{a^2} : \quad \mathbf{w}_{4,O} = [0 \quad 0 \quad 0 \quad 0 \quad 0 \quad 1]^{\text{T}},$$

$$\lambda_5 = 0.00352\frac{EA}{a^2} : \quad \mathbf{w}_{5,O} = [0 \quad 0 \quad 0 \quad 0 \quad 1 \quad 0]^{\text{T}},$$

$$\lambda_6 = 0 : \quad \mathbf{w}_{6,O} = [0 \quad 0 \quad 0 \quad 1 \quad 0 \quad 0]^{\text{T}}.$$

$$(5.13)$$

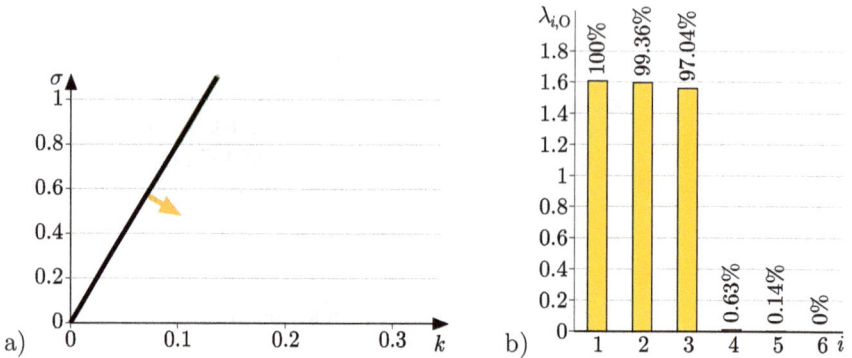

Figure 5.16 Extremal properties of the regular expanded octahedron: a) line of extremal properties $\sigma = 8.034 \cdot k$; b) distribution of eigenvalues for $k_{\text{extr}} = 0.1$, $\sigma_{\text{extr}} = 0.8034$ (multiplier EA/a^2).

Extremal properties of the module are presented in Figure 5.16.

The module can be identified as trimode, since one eigenvalue is equal to zero and two others are smaller than 1% of the maximum eigenvalue. Three positive eigenvalues have similar values. Analysis of the graphs (Figure 5.16) leads to the conclusion that the regular expanded octahedron has three stiff and three soft (one zero eigenvalue and two eigenvalues around 0.1% and 0.7% of $\lambda_{\text{max,O}}$) modes of deformation.

The stiff mode (Figure 5.17a), represented by the eigenvector $\mathbf{w}_{1,\text{O}}$, is volumetric with the dominant extension in X_1 direction. The stiff mode (Figure 5.17b), represented by the eigenvector $\mathbf{w}_{2,\text{O}}$, is volumetric with the dominant extension in X_2 direction. The stiff mode (Figure 5.17c), represented by the eigenvector $\mathbf{w}_{3,\text{O}}$, is volumetric with the dominant extension in X_3 direction. The soft mode (Figure 5.17d), represented by the eigenvector $\mathbf{w}_{4,\text{O}}$, is a shear deformation in X_2–X_3 plane. The soft mode (Figure 5.17e), represented by the eigenvector $\mathbf{w}_{5,\text{O}}$, is a shear deformation in X_1–X_3 plane. The soft mode (Figure 5.17f), represented by the eigenvector $\mathbf{w}_{6,\text{O}}$, is a shear deformation in X_1–X_2 plane.

Apart from the regular module inscribed into a cube of edge length a, four other variants of the expanded octahedron module were analysed (Figure 5.18), with the proportions described in Section 5.2.

The elasticity matrices obtained for the considered variants of the expanded octahedron have the same form as the elasticity matrix of the regular module $\mathbf{E}_{\text{O}}$ (Eq. 5.11), but with different coefficients e_{ij}.

Figure 5.19 presents distributions of eigenvalues obtained for five variants of the expanded octahedron. For each eigenvalue a percentage value is given— the eigenvalue corresponding to the stiff mode is assumed as a reference value of 100% and for the following modes a ratio of $\lambda_{i,\text{O}}/\lambda_{\text{max,O}} \cdot 100\%$ is calculated. The percentage values marked with colours indicate soft modes of

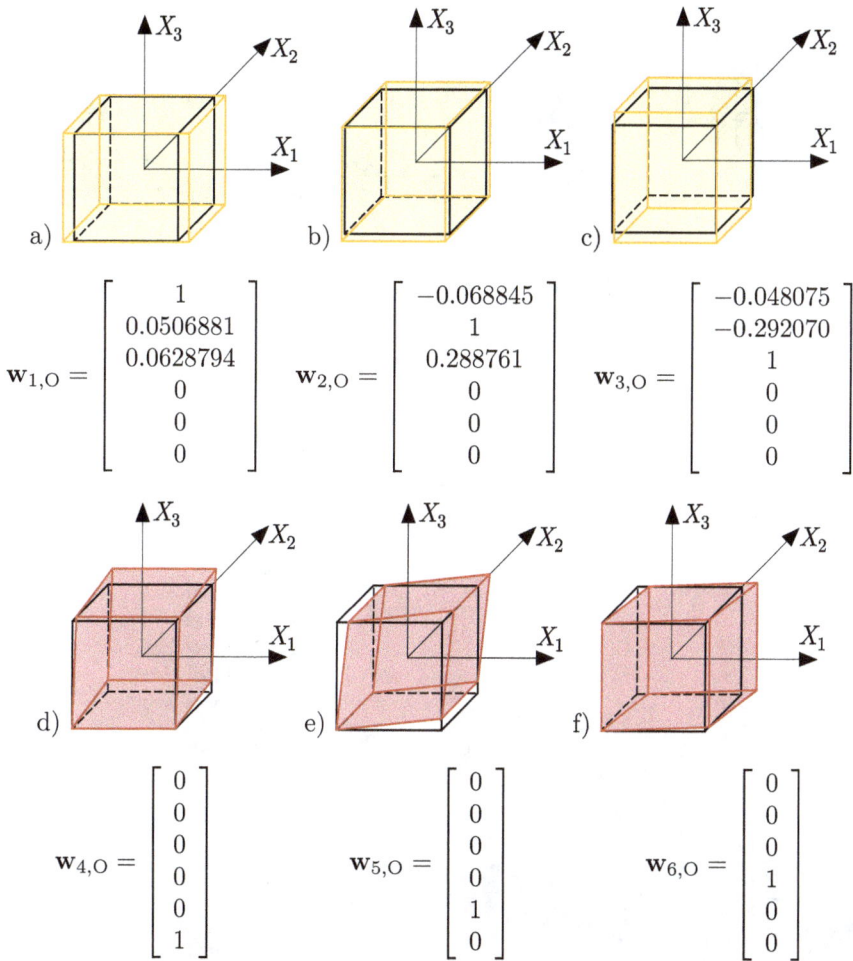

$$\mathbf{w}_{1,O} = \begin{bmatrix} 1 \\ 0.0506881 \\ 0.0628794 \\ 0 \\ 0 \\ 0 \end{bmatrix} \qquad \mathbf{w}_{2,O} = \begin{bmatrix} -0.068845 \\ 1 \\ 0.288761 \\ 0 \\ 0 \\ 0 \end{bmatrix} \qquad \mathbf{w}_{3,O} = \begin{bmatrix} -0.048075 \\ -0.292070 \\ 1 \\ 0 \\ 0 \\ 0 \end{bmatrix}$$

$$\mathbf{w}_{4,O} = \begin{bmatrix} 0 \\ 0 \\ 0 \\ 0 \\ 0 \\ 1 \end{bmatrix} \qquad \mathbf{w}_{5,O} = \begin{bmatrix} 0 \\ 0 \\ 0 \\ 0 \\ 1 \\ 0 \end{bmatrix} \qquad \mathbf{w}_{6,O} = \begin{bmatrix} 0 \\ 0 \\ 0 \\ 1 \\ 0 \\ 0 \end{bmatrix}$$

Figure 5.17 Deformation modes of the regular expanded octahedron: a) stiff mode corresponding to λ_1; b) stiff mode corresponding to λ_2; c) stiff mode corresponding to λ_3; d) soft mode corresponding to λ_4; e) soft mode corresponding to λ_5; f) soft mode corresponding to λ_6.

deformation (the colours correspond to the module variants). Extremal properties presented in Figure 5.19 were obtained for the parameters $k_{\text{extr}} = 0.1$ and σ_{extr} given in Table 5.6, depending on the analysed variant.

All variants of the expanded octahedron can be identified as trimode, since in each case, three eigenvalues are smaller than 1% of the maximum eigenvalue obtained for the given variant. While the regular module has three stiff modes of deformation, two variants (low and very low) have two stiff and one medium mode, and two others (high and very high) have one stiff and

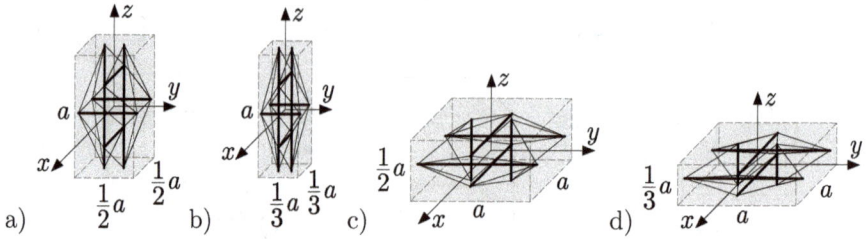

Figure 5.18 Variants of the expanded octahedron module: a) high; b) very high; c) low; d) very low.

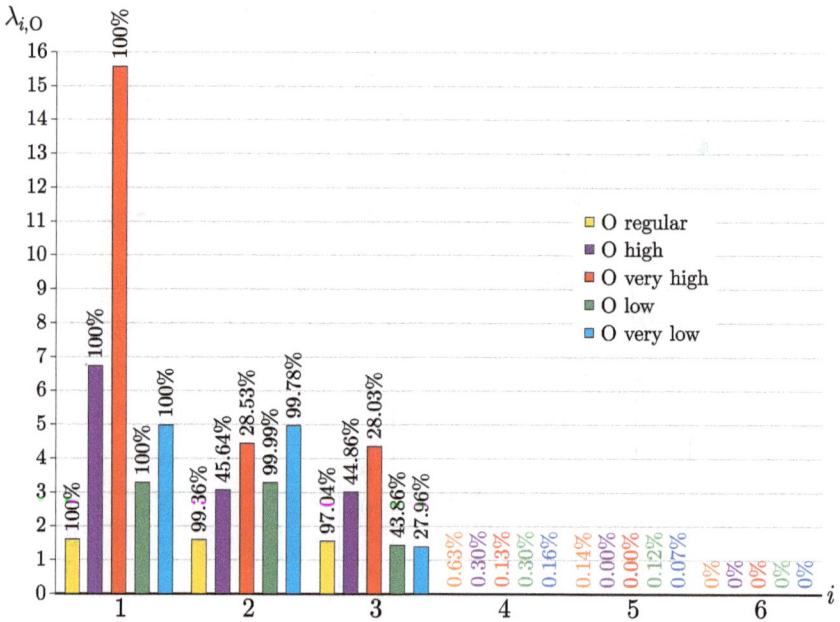

Figure 5.19 Distribution of eigenvalues for five variants of the expanded octahedron module (multiplier EA/a^2).

Table 5.6

Extremal properties of five expanded octahedron variants ($k_{\text{extr}} = 0.1$).

Module	σ_{extr}	Extremal Properties
regular	0.8034	trimode
high	0.7338	trimode
very high	0.7240	trimode
low	0.9000	trimode
very low	0.9373	trimode

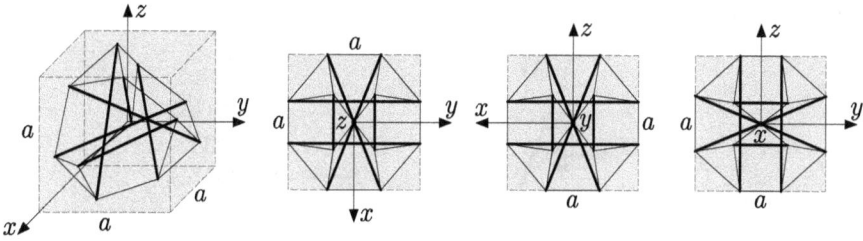

Figure 5.20 Regular truncated tetrahedron module inscribed into a cube.

two medium modes. The variant denoted as very high exhibits the biggest differences between the eigenvalue corresponding to the stiff mode and the ones indicating medium modes of deformation (eigenvalues around 28% of $\lambda_{\max,O}$).

5.2.4 TRUNCATED TETRAHEDRON FAMILY

The analysed regular truncated tetrahedron module (T) is presented in Figure 5.20. Geometry of the module is described in Table 5.7, which contains nodal coordinates of the truncated tetrahedron.

Table 5.7

Geometry of the regular truncated tetrahedron.

Node No.	x	y	z	Scheme
1	$0.188 \cdot a$	$0.145 \cdot a$	$0.500 \cdot a$	
2	$0.145 \cdot a$	$0.500 \cdot a$	$0.188 \cdot a$	
3	$0.500 \cdot a$	$0.188 \cdot a$	$0.145 \cdot a$	
4	$-0.188 \cdot a$	$-0.145 \cdot a$	$0.500 \cdot a$	
5	$-0.145 \cdot a$	$-0.500 \cdot a$	$0.188 \cdot a$	
6	$-0.500 \cdot a$	$-0.188 \cdot a$	$0.145 \cdot a$	
7	$-0.500 \cdot a$	$0.188 \cdot a$	$-0.145 \cdot a$	
8	$-0.188 \cdot a$	$0.145 \cdot a$	$-0.500 \cdot a$	
9	$-0.145 \cdot a$	$0.500 \cdot a$	$-0.188 \cdot a$	
10	$0.500 \cdot a$	$-0.188 \cdot a$	$-0.145 \cdot a$	
11	$0.188 \cdot a$	$-0.145 \cdot a$	$-0.500 \cdot a$	
12	$0.145 \cdot a$	$-0.500 \cdot a$	$-0.188 \cdot a$	

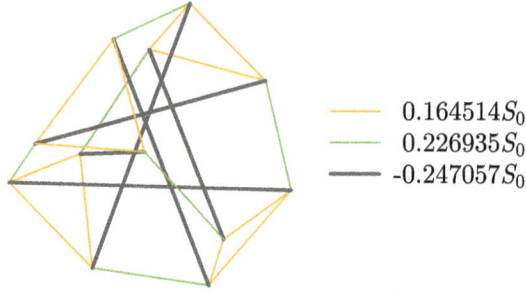

Figure 5.21 Self-stress state in the regular truncated tetrahedron.

The module has seven infinitesimal mechanisms and one self-stress state that eliminates them (Figure 5.21)—self-stress is expressed by relative forces in struts and cables with a multiplier S_0.

The elasticity matrix obtained from the continuum model has a form

$$
\mathbf{E_T} =
\begin{bmatrix}
e_{11} & e_{12} & e_{12} & 0 & 0 & 0 \\
 & e_{11} & e_{12} & 0 & 0 & 0 \\
 & & e_{11} & 0 & 0 & 0 \\
 & & & e_{12} & 0 & 0 \\
 & & & & e_{12} & 0 \\
\text{sym.} & & & & & e_{12}
\end{bmatrix},
\tag{5.14}
$$

$$
e_{11} = \frac{2EA}{a^2}(0.836131 + 0.728014 \cdot k + 0.0709611 \cdot \sigma),
$$
$$
e_{12} = \frac{EA}{a^2}(0.232405 + 0.696312 \cdot k - 0.0709611 \cdot \sigma).
\tag{5.15}
$$

Similarly to the previous module, the regular truncated tetrahedron has orthotropic properties. Spectral analysis of the elasticity matrix allows us to find eigenvalues and corresponding eigenvectors, which describe extremal properties of the module. The identification of extremal mechanical behaviour was performed in the same way as described in Section 5.2.1. The following eigenvalues and corresponding eigenvectors were obtained for $k_{\text{extr}} = 0.1$ and $\sigma_{\text{extr}} = 4.2563$:

$$
\lambda_1 = 2.42194\frac{EA}{a^2} : \quad \mathbf{w}_{1,\mathrm{T}} = [1 \ \ 1 \ \ 1 \ \ 0 \ \ 0 \ \ 0]^{\mathrm{T}},
$$
$$
\lambda_2 = 2.42186\frac{EA}{a^2} : \quad \mathbf{w}_{2,\mathrm{T}} = [1 \ \ -1 \ \ 0 \ \ 0 \ \ 0 \ \ 0]^{\mathrm{T}},
$$
$$
\lambda_3 = 2.42186\frac{EA}{a^2} : \quad \mathbf{w}_{3,\mathrm{T}} = [-0.5 \ \ -0.5 \ \ 1 \ \ 0 \ \ 0 \ \ 0]^{\mathrm{T}}, \tag{5.16}
$$
$$
\lambda_4 = 0 : \quad \mathbf{w}_{4,\mathrm{T}} = [0 \ \ 0 \ \ 0 \ \ 1 \ \ 0 \ \ 0]^{\mathrm{T}},
$$
$$
\lambda_5 = 0 : \quad \mathbf{w}_{5,\mathrm{T}} = [0 \ \ 0 \ \ 0 \ \ 0 \ \ 1 \ \ 0]^{\mathrm{T}},
$$
$$
\lambda_6 = 0 : \quad \mathbf{w}_{6,\mathrm{T}} = [0 \ \ 0 \ \ 0 \ \ 0 \ \ 0 \ \ 1]^{\mathrm{T}}.
$$

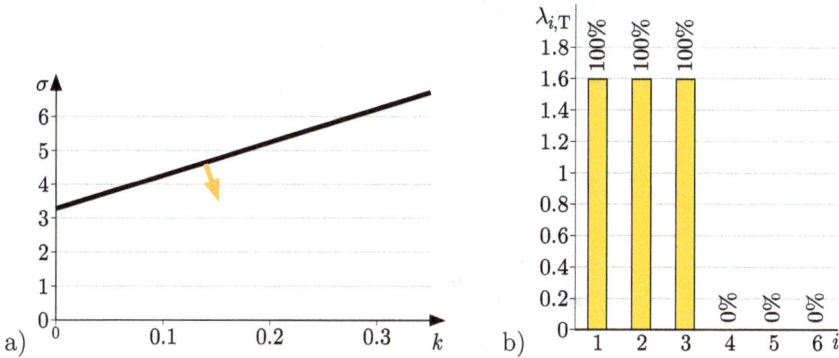

Figure 5.22 Extremal properties of the regular truncated tetrahedron: a) line of extremal properties $\sigma = 3.275 + 9.81 \cdot k$; b) distribution of eigenvalues for $k_{\text{extr}} = 0.1$, $\sigma_{\text{extr}} = 4.2563$ (multiplier EA/a^2).

Extremal properties of the module are presented in Figure 5.22.

The module can be identified as trimode, since three eigenvalues are equal to zero. Three positive eigenvalues are equal. Unfortunately, the values of σ are unreal as far as the currently available technology is concerned.

Analysis of the graphs (Figure 5.22) leads to the conclusion that the regular truncated tetrahedron has three stiff and three purely soft modes of deformation. It is the most extreme structure from all analysed examples of regular modules. However, realisation of such a structure is not possible with the available materials.

The stiff mode (Figure 5.23a), represented by the eigenvector $\mathbf{w}_{1,\text{T}}$, is volumetric with the uniform extension in all directions. The stiff mode (Figure 5.23b), represented by the eigenvector $\mathbf{w}_{2,\text{T}}$, is volumetric with the extension in X_1 direction equal to the contraction in X_2 direction. The stiff mode (Figure 5.23c), represented by the eigenvector $\mathbf{w}_{3,\text{T}}$, is volumetric with the extension in X_3 direction two times bigger than the contractions in two other directions. The soft mode (Figure 5.23d), represented by the eigenvector $\mathbf{w}_{4,\text{T}}$, is a shear deformation in X_1–X_2 plane. The soft mode (Figure 5.23e), represented by the eigenvector $\mathbf{w}_{5,\text{T}}$, is a shear deformation in X_1–X_3 plane. The soft mode (Figure 5.23f), represented by the eigenvector $\mathbf{w}_{6,\text{T}}$, is a shear deformation in X_2–X_3 plane.

Apart from the regular module inscribed into a cube of edge length a, four other variants of the truncated tetrahedron module were analysed (Figure 5.24), with the proportions described in Section 5.2.

The elasticity matrices obtained for the considered variants of the truncated tetrahedron have the same form as the elasticity matrix of the regular module $\mathbf{E}_\text{T}$ (Eq. 5.14), but with different coefficients e_{ij}.

Figure 5.25 presents distributions of eigenvalues obtained for five variants of the truncated tetrahedron. For each eigenvalue a percentage value is

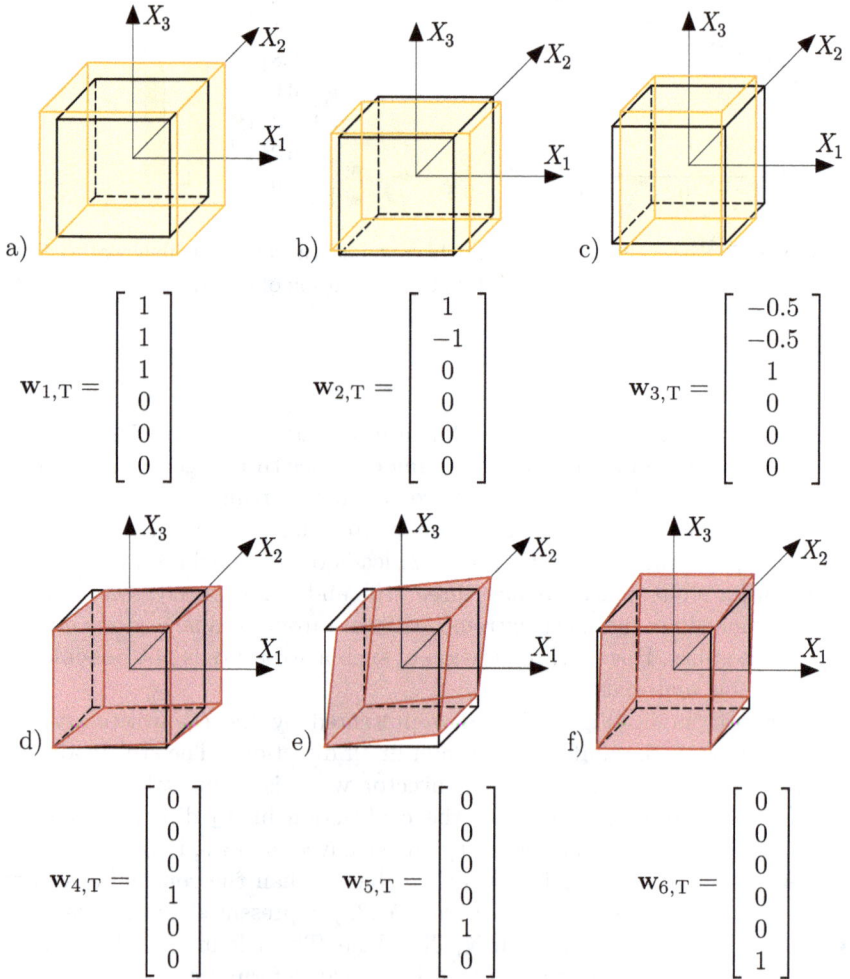

Figure 5.23 Deformation modes of the regular truncated tetrahedron: a) stiff mode corresponding to λ_1; b) stiff mode corresponding to λ_2; c) stiff mode corresponding to λ_3; d) soft mode corresponding to λ_4; e) soft mode corresponding to λ_5; f) soft mode corresponding to λ_6.

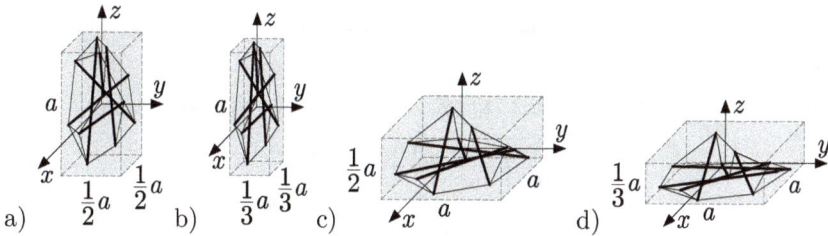

Figure 5.24 Variants of the truncated tetrahedron module: a) high; b) very high; c) low; d) very low.

given—the eigenvalue corresponding to the stiff mode is assumed as a reference value of 100% and for the following modes a ratio of $\lambda_{i,T}/\lambda_{\max,T} \cdot 100\%$ is calculated. The percentage values marked with colours indicate soft modes of deformation (the colours correspond to the module variants). Extremal properties presented in Figure 5.25 were obtained for the parameters $k_{\mathrm{extr}} = 0.1$ and σ_{extr} given in Table 5.8, depending on the analysed variant.

While the regular truncated tetrahedron is trimode, other variants of this module can be identified as unimode, since in each case, there is only one eigenvalue equal to zero. Moreover, stiff modes of deformation differ between

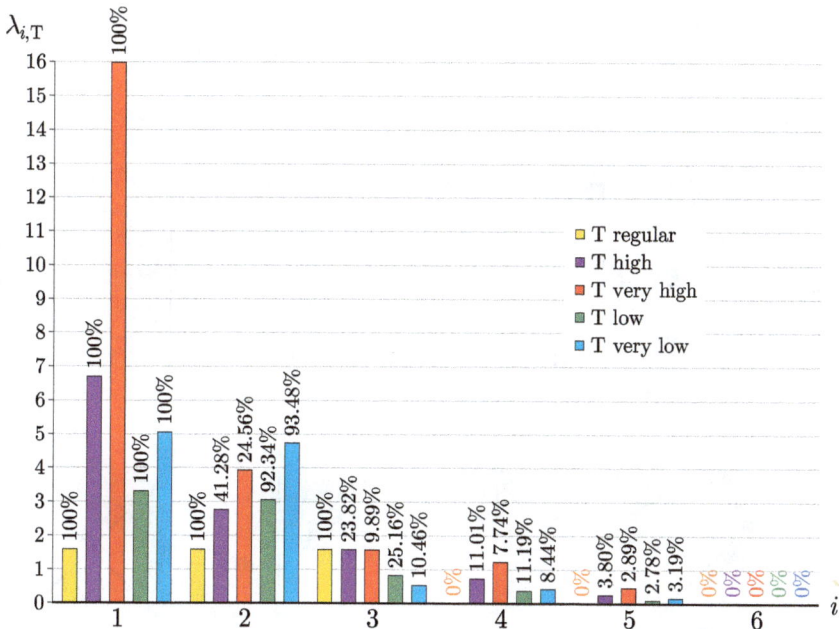

Figure 5.25 Distribution of eigenvalues for five variants of the truncated tetrahedron module (multiplier EA/a^2).

Table 5.8

Extremal properties of five truncated tetrahedron variants ($k_{extr} = 0.1$).

Module	σ_{extr}	Extremal Properties
regular	4.2563	trimode
high	2.2181	unimode
very high	1.8743	unimode
low	2.7817	unimode
very low	2.5454	unimode

the variants. The regular module has three stiff modes, two variants (low and very low) have two stiff and three medium modes, and two others (high and very high) have one stiff and four medium modes of deformation. Truncated tetrahedron is an example of the module, where the change of geometrical proportions leads to a significantly different extremal behaviour.

5.2.5   X-MODULE FAMILY

The analysed regular X-module (X) is presented in Figure 5.26. Geometry of the module is described in Table 5.9, which contains nodal coordinates of the X-module.

The module has one infinitesimal mechanism and one corresponding self-stress state (Figure 5.27)—self-stress is expressed by relative forces in struts and cables with a multiplier S_0.

The elasticity matrix obtained from the continuum model has a form

$$
\mathbf{E_X} =
\begin{bmatrix}
e_{11} & e_{12} & e_{13} & 0 & e_{15} & 0 \\
 & e_{22} & e_{23} & 0 & -e_{23} & 0 \\
 & & e_{33} & 0 & e_{15} & 0 \\
 & & & e_{12} & 0 & -e_{23} \\
 & & & & e_{13} & 0 \\
\text{sym.} & & & & & e_{23}
\end{bmatrix},
\tag{5.17}
$$

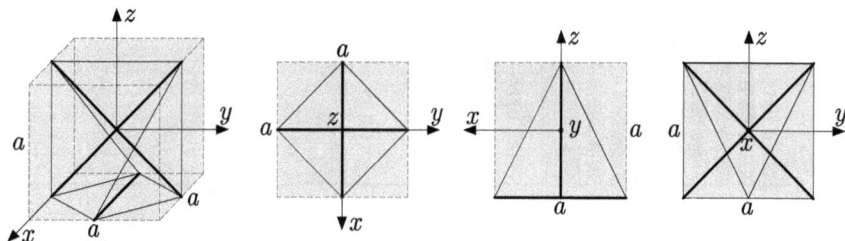

Figure 5.26   Regular X-module inscribed into a cube.

Table 5.9
Geometry of the regular X-module.

Node No.	x	y	z	Scheme
1	0	$-0.5 \cdot a$	$-0.5 \cdot a$	
2	$0.5 \cdot a$	0	$-0.5 \cdot a$	
3	0	$0.5 \cdot a$	$-0.5 \cdot a$	
4	$-0.5 \cdot a$	0	$-0.5 \cdot a$	
5	0	$-0.5 \cdot a$	$0.5 \cdot a$	
6	0	$0.5 \cdot a$	$0.5 \cdot a$	

$$e_{11} = \frac{2EA}{a^2}(0.353553 + 0.887574 \cdot k - 0.0790569 \cdot \sigma),$$

$$e_{12} = \frac{EA}{a^2}(0.707107 + 0.272166 \cdot k + 0.316228 \cdot \sigma),$$

$$e_{13} = \frac{EA}{a^2}(0.775148 \cdot k - 0.158114 \cdot \sigma),$$

$$e_{15} = \frac{EA}{a^2}(-0.0680414 \cdot k),$$

$$e_{22} = \frac{2EA}{a^2}(0.353553 + 1.54433 \cdot k - 0.158114 \cdot \sigma),$$

$$e_{23} = \frac{EA}{a^2}(0.272166 \cdot k),$$

$$e_{33} = \frac{2EA}{a^2}(0.5 + 0.387574 \cdot k + 0.0790569 \cdot \sigma).$$

(5.18)

The elasticity matrix indicates that the regular X-module has anisotropic properties. Spectral analysis of this matrix allows us to find eigenvalues and corresponding eigenvectors, which describe extremal properties of the

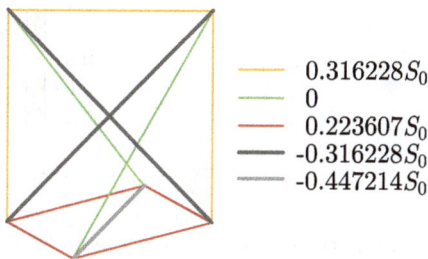

	$0.316228 S_0$
	0
	$0.223607 S_0$
	$-0.316228 S_0$
	$-0.447214 S_0$

Figure 5.27 Self-stress state in the regular X-module.

module. The identification of extremal mechanical behaviour was performed in the same way as described in Section 5.2.1. The following eigenvalues and corresponding eigenvectors were obtained for $k_{\text{extr}} = 0.1$ and $\sigma_{\text{extr}} = 0.3700$:

$$\lambda_1 = 1.71686\frac{EA}{a^2} : \quad \mathbf{w}_{1,\text{O}} = [0.957577 \quad 1 \quad 0.078437 \quad 0 \quad -0.020182 \quad 0]^{\text{T}},$$

$$\lambda_2 = 1.13421\frac{EA}{a^2} : \quad \mathbf{w}_{2,\text{O}} = [-0.042566 \quad -0.037777 \quad 1 \quad 0 \quad -0.004929 \quad 0]^{\text{T}},$$

$$\lambda_3 = 0.85223\frac{EA}{a^2} : \quad \mathbf{w}_{3,\text{O}} = [0 \quad 0 \quad 0 \quad 1 \quad 0 \quad -0.032989]^{\text{T}},$$

$$\lambda_4 = 0.02900\frac{EA}{a^2} : \quad \mathbf{w}_{4,\text{O}} = [0.549374 \quad -0.506605 \quad 0.009166 \quad 0 \quad 1 \quad 0]^{\text{T}},$$

$$\lambda_5 = 0.02632\frac{EA}{a^2} : \quad \mathbf{w}_{5,\text{O}} = [0 \quad 0 \quad 0 \quad 0.032989 \quad 0 \quad 1]^{\text{T}},$$

$$\lambda_6 = 0 : \quad \mathbf{w}_{6,\text{O}} = [-0.956738 \quad 0.936366 \quad -0.000432 \quad 0 \quad 1 \quad 0]^{\text{T}}.$$

$$(5.19)$$

Extremal properties of the module are presented in Figure 5.28.

The module can be identified as quasi trimode, since one eigenvalue is equal to zero and two others are smaller than 2% of the maximum eigenvalue. One positive eigenvalue is dominant over the others.

Analysis of the graphs (Figure 5.28) leads to the conclusion that the regular X-module has one stiff, three soft (one purely soft with the corresponding zero eigenvalue, and two quasi soft with the eigenvalues around 1.5% and 1.7% of $\lambda_{\text{max},\text{X}}$) and two medium (the eigenvalues around 49% and 66% of $\lambda_{\text{max},\text{X}}$) modes of deformation.

The stiff mode (Figure 5.29a), represented by the eigenvector $\mathbf{w}_{1,\text{X}}$, is a combination of a volumetric deformation with the dominant extension in X_1 and X_2 directions and a small shear deformation in X_1–X_3 plane. The quasi

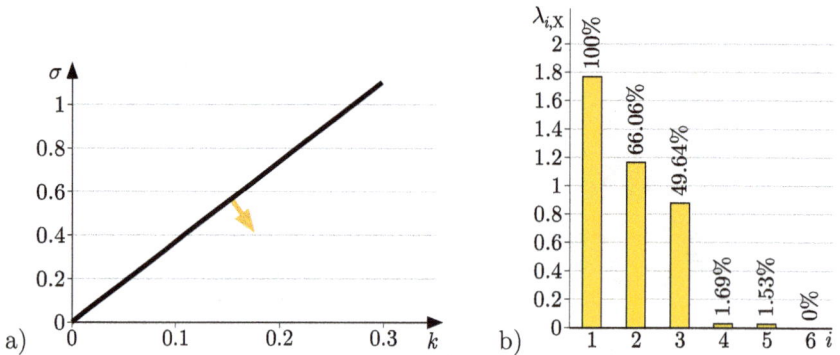

Figure 5.28 Extremal properties of the regular X-module: a) line of extremal properties $\sigma = 0.001 + 3.69 \cdot k$; b) distribution of eigenvalues for $k_{\text{extr}} = 0.1$, $\sigma_{\text{extr}} = 0.3700$ (multiplier EA/a^2).

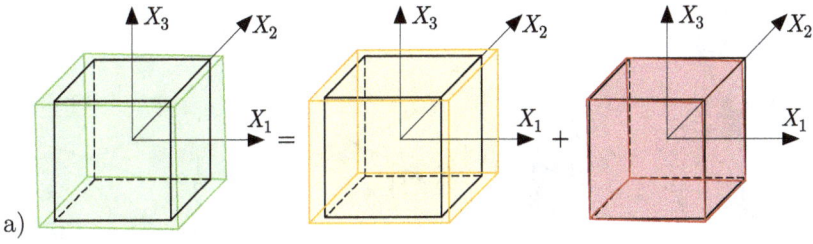

$$\mathbf{w}_{1,\mathrm{X}} = [0.957577 \quad 1 \quad 0.078437 \quad 0 \quad -0.020182 \quad 0]^{\mathrm{T}}$$

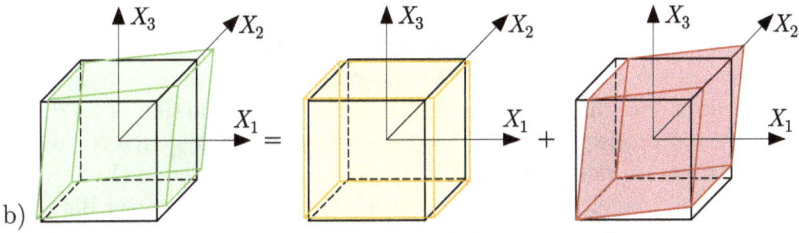

$$\mathbf{w}_{4,\mathrm{X}} = [0.549374 \quad -0.506605 \quad 0.009166 \quad 0 \quad 1 \quad 0]^{\mathrm{T}}$$

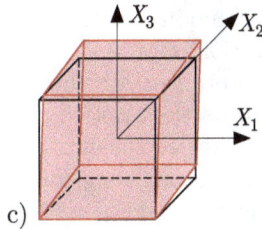

$$\mathbf{w}_{5,\mathrm{X}} = [0 \quad 0 \quad 0 \quad 0.032989 \quad 0 \quad 1]^{\mathrm{T}}$$

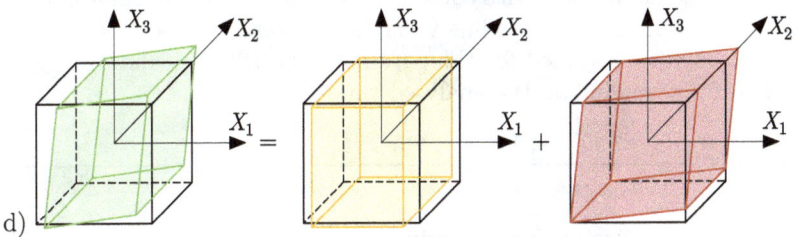

$$\mathbf{w}_{6,\mathrm{X}} = [-0.956738 \quad 0.936366 \quad -0.000432 \quad 0 \quad 1 \quad 0]^{\mathrm{T}}$$

Figure 5.29 Deformation modes of the regular X-module: a) stiff mode corresponding to λ_1; b) quasi soft mode corresponding to λ_4; c) quasi soft mode corresponding to λ_5; d) soft mode corresponding to λ_6.

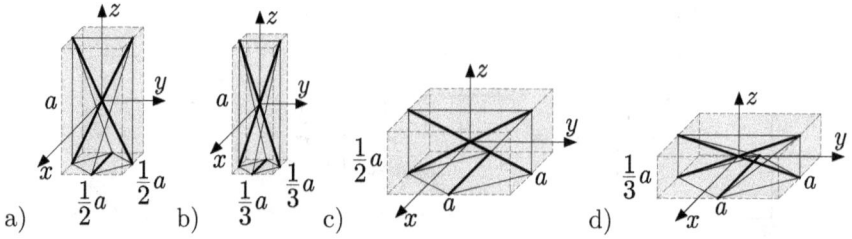

Figure 5.30 Variants of the X-module: a) high; b) very high; c) low; d) very low.

soft mode (Figure 5.29b), represented by the eigenvector $\mathbf{w}_{4,X}$, is a combination of a shear deformation in X_1–X_3 plane and a volumetric deformation with two dominant directions: extension in X_1 and contraction in X_2 direction. The quasi soft mode (Figure 5.29c), represented by the eigenvector $\mathbf{w}_{5,X}$, is a shear deformation dominant in X_2–X_3 plane. The soft mode (Figure 5.29d), represented by the eigenvector $\mathbf{w}_{6,X}$, is a combination of a shear deformation in X_1–X_3 plane and a volumetric deformation with two dominant directions: contraction in X_1 and extension in X_2 direction.

Apart from the regular module inscribed into a cube of edge length a, four other variants of the X-module were analysed (Figure 5.30), with the proportions described in Section 5.2.

The elasticity matrices obtained for the considered variants of the X-module have the same form as the elasticity matrix of the regular module $\mathbf{E}_X$ (Eq. 5.17), but with different coefficients e_{ij}.

Figure 5.31 presents distributions of eigenvalues obtained for five variants of the X-module. For each eigenvalue a percentage value is given—the eigenvalue corresponding to the stiff mode is assumed as a reference value of 100% and for the following modes a ratio of $\lambda_{i,X}/\lambda_{\max,X} \cdot 100\%$ is calculated. The percentage values marked with colours indicate soft modes of deformation (the colours correspond to the module variants). Extremal properties presented in Figure 5.31 were obtained for the parameters $k_{\text{extr}} = 0.1$ and σ_{extr} given in Table 5.10, depending on the analysed variant.

Table 5.10

Extremal properties of five X-module variants ($k_{\text{extr}} = 0.1$).

Module	σ_{extr}	Extremal Properties
regular	0.3700	quasi trimode
high	0.4828	trimode
very high	0.6775	trimode
low	0.4310	quasi trimode
very low	0.4742	quasi trimode

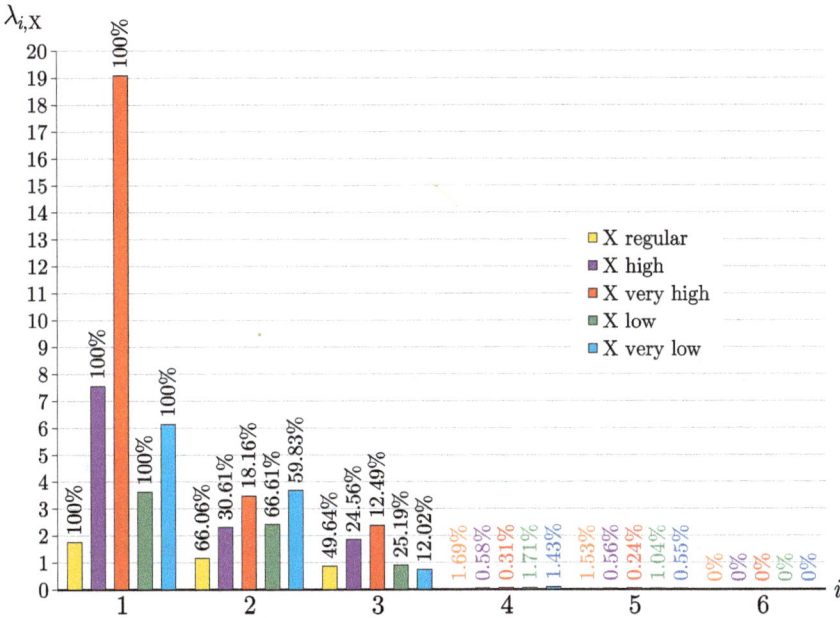

Figure 5.31 Distribution of eigenvalues for five variants of the X-module (multiplier EA/a^2).

Three variants of the X-module (regular, low and very low) can be identified as quasi trimode and two others (high and very high) as trimode, since in these cases, all three eigenvalues are smaller than 1% of the maximum eigenvalue obtained for the given variant. The variant denoted as very high exhibits the biggest differences between the eigenvalue corresponding to the stiff mode and the ones indicating medium modes of deformation (eigenvalues around 18% and 12% of $\lambda_{\text{max,X}}$).

5.2.6 PARAMETRIC ANALYSIS

Structural properties of tensegrity systems can be controlled using various parameters. In Section 3.3 it was proved that it is possible to change stiffness of tensegrity structures by adjusting the level of self-stress (which can be expressed by $\sigma = S_0/EA$) and this unique feature allows us to qualify tensegrities to smart structures. Another parameter that influences structural behaviour is the ratio of cable to strut stiffness $k = (EA)_{\text{cable}}/(EA)_{\text{strut}}$ defined in Section 5.2. Both parameters k and σ have a significant influence on extremal properties of tensegrity structures. However, from the practical point of view, only one of these parameters can be adjusted after the structure has been constructed. Parameter k depends on applied materials and cross-sections of structural members and it would be hard to change such elements

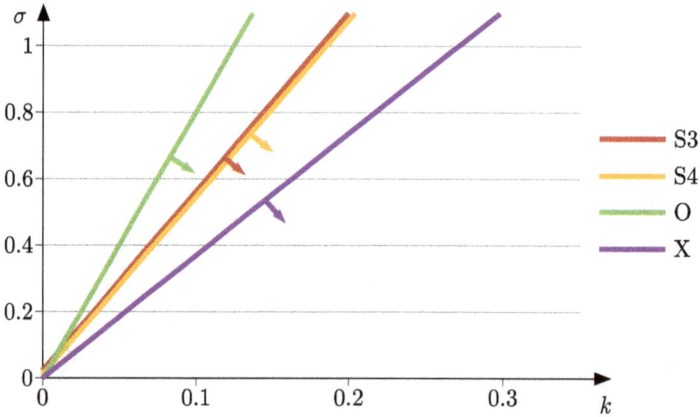

Figure 5.32 Lines of extremal properties for selected regular tensegrity modules.

in an existing structure. Parameter σ, on the other hand, is correlated with the level of prestressing forces and this is something which can be quite easily adjusted in an operating system by using actuators.

The influence of parameters k and σ on structural properties of selected tensegrity systems is shown in Figure 5.32, which depicts lines of extremal mechanical properties obtained for four out of five analysed regular modules. As explained earlier, the line determined for the regular truncated tetrahedron goes beyond the range of achievable values of parameter σ.

The biggest area where the elasticity matrix $\mathbf{E}$ is positive definite (Figure 5.32) is observed for the regular expanded octahedron (O) and the smallest—for the regular X-module (X). The areas for both regular simplex modules (S3 and S4) are similar. Moreover, it can be noticed that the eigenvalues for stiff modes that accompany the identified soft modes (Figures 5.4b, 5.10b, 5.16b, 5.22b and 5.28b) are on the same level for all regular modules. Soft deformation modes are volumetric with various signs for the regular three-strut simplex, four-strut simplex and expanded octahedron, but shear dominated for the truncated tetrahedron and the X-module. Stiff modes of deformation are more or less uniform volumetric for the regular three-strut simplex, four-strut simplex and truncated tetrahedron, volumetric in two directions for the X-module and volumetric in one direction for the expanded octahedron.

The study presented in Sections 5.2.1–5.2.5 revealed that apart from material properties, cross-sections of structural members and self-stress level, extremal behaviour of tensegrity systems depends on geometrical proportions of the modules. Figure 5.33 depicts lines of extremal mechanical properties obtained for the variants of four out of five analysed tensegrity modules.

Analysis of these graphs (Figure 5.33) leads to the conclusion that there is no repeating pattern in the correlation between the geometrical proportions

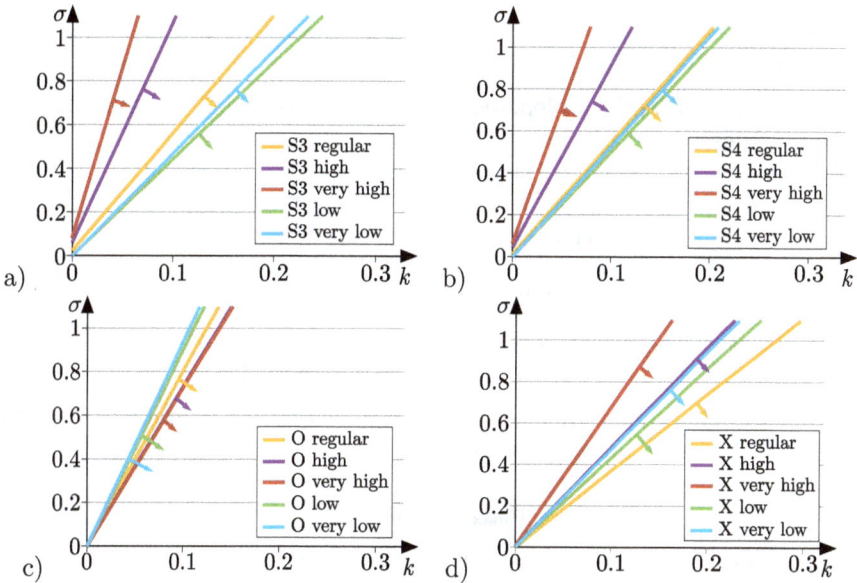

Figure 5.33 Lines of extremal properties for five variants of: a) three-strut simplex; b) four-strut simplex; c) expanded octahedron; d) X-module.

and the parameters for which soft modes of deformation are obtained. By constructing lower variants of the simplex modules, lower values of σ_{extr} are needed to obtain extremal behaviour of the structure. However, in the case of the X-module, the most favourable values of σ_{extr} are obtained for the regular module, and in the case of the expanded octahedron—for its very high variant.

From the practical point of view, in order to analyse extremal properties of structures (for example, the graphs presented in Figures 5.32 and 5.33), a discussion on material parameters is needed. For this purpose, let us consider a tensegrity module constructed from real structural elements (cables and struts) made of currently available materials. It is assumed that the module is inscribed into a cube of edge length $a = 1$ m and typical elements are used for structural members:

- cables—wires with a diameter of 1 mm, made of titanium alloy Titan Grade 5 / 6Al4V with tensile strength of 1400 MPa and Young's modulus of 120 GPa [134];
- struts—hollow cross-section with a diameter of 54 mm and a wall thickness of 2.6 mm, made of steel S355 with a load-carrying capacity of 84 kN [65] and Young's modulus of 210 GPa.

Parameter k can be calculated as

$$k = \frac{(EA)_{\text{cable}}}{(EA)_{\text{strut}}} = \frac{120 \text{ GPa} \cdot 0.785 \text{ mm}^2}{210 \text{ GPa} \cdot 419.84 \text{ mm}^2} = 0.00107. \tag{5.20}$$

Table 5.11

Parameters k_{extr} and σ_{extr} leading to extremal properties of regular tensegrity modules.

Module	k_{extr}	σ_{extr}
three-strut simplex	0.00107	0.00620
four-strut simplex	0.00107	0.00593
expanded octahedron	0.00107	0.00859
truncated tetrahedron	0.00107	3.285
X-module	0.00107	0.00396

Taking into account the load-carrying capacity of structural members, the maximum value of parameter σ is

$$\sigma_{\text{max}} = 0.0000374. \qquad (5.21)$$

This value is much smaller than the values that are needed to obtain soft modes of deformation in the analysed modules. Table 5.11 contains the values of σ_{extr}, for which the regular modules exhibit extremal properties, with an assumption that $k_{\text{extr}} = 0.00107$.

Although the values in Table 5.11 are not achievable with the assumed materials, some of these modules can exhibit extremal behaviour using the adopted typical members. Figure 5.34 presents distribution of eigenvalues obtained for five regular tensegrity modules, for real parameters calculated above: $k = 0.00107$ and $\sigma_{\text{max}} = 0.0000374$. For each eigenvalue a percentage value is given—the eigenvalue corresponding to the stiff mode is assumed as a reference value of 100% and for the following modes a ratio of $\lambda_i / \lambda_{\text{max}} \cdot 100\%$ is calculated. The percentage values marked with colours indicate soft modes of deformation (the colours correspond to regular modules). Here, the eigenvalues are not scaled as it is assumed that all modules are constructed from the same struts and cables.

It can be noticed that four out of five considered modules exhibit extremal properties: three-strut simplex, expanded octahedron and X-module may be described as trimode, four-strut simplex as bimode. The truncated tetrahedron exhibits no extremal behaviour as the difference between σ_{max} and σ that is needed to obtain soft modes of deformation in this module is too big. The smallest eigenvalues obtained in this example, contrary to the examples presented in previous sections, do not exactly equal zero. They are however very small compared to the biggest eigenvalues (less than 1% of λ_{max}) and therefore, it can be assumed that they indicate soft modes of deformation.

Sections 5.2.1–5.2.5 were focussed on a systematic study on extremal mechanical properties of a series of 3D tensegrity modules. All presented analyses were based on the continuum approach presented in Section 4.2, which can be

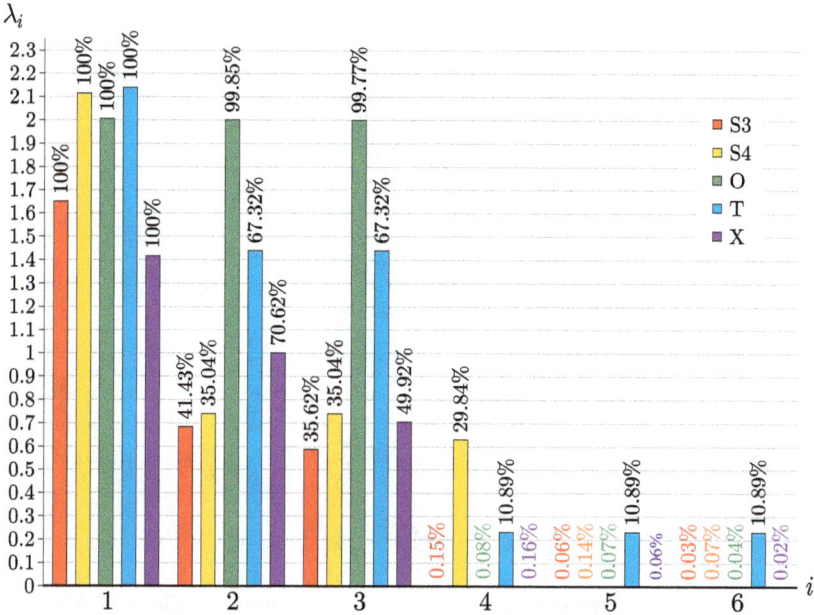

Figure 5.34 Distribution of eigenvalues for regular modules for $k = 0.00107$, $\sigma_{max} = 0.0000374$ (multiplier EA/a^2).

used for the identification of extremal behaviour of metamaterials. The aim of Section 5.2.6, on the other hand, was to assess theoretical results obtained for five tensegrity modules from the engineering point of view. It was demonstrated that although the theoretically determined values of parameters k_{extr} and σ_{extr} are not achievable using the currently existing technology, some of these modules can exhibit extremal behaviour with commonly available parent materials. Moreover, looking at the rapid development of material sciences and modern technologies, such as for example the additive manufacturing, it is safe to assume that current limitations in the construction of metamaterials of certain parameters will soon be overcome.

5.3 TENSEGRITY LATTICES IN 3D SPACE

Tensegrity modules described in Section 5.2 can be used to create more complex structures, such as tensegrity lattices. There are three ways in which such systems can be constructed:

- modular system—the structure is composed of regular tensegrity modules joined in nodes without any additional elements (Figure 5.35a),
- non-modular system—the structure is based on regular modules, but they are connected in such a way that they create a new tensegrity system (Figure 5.35b),

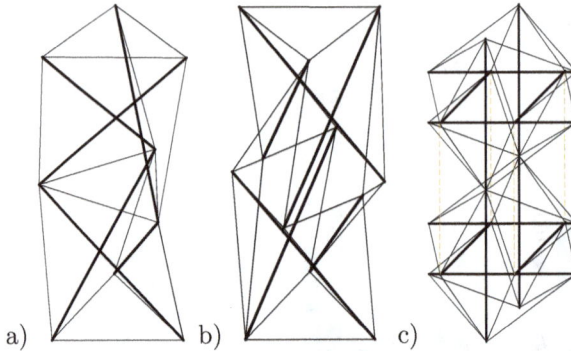

Figure 5.35 Connection of tensegrity modules: a) three-strut simplex modules connected in nodes; b) three-strut simplex modules connected between nodes; c) expanded octahedron modules connected in nodes, with extra cables (dashed lines).

- modular system with extra cables—the structure is created similarly to the modular system, but some additional cables connecting adjacent modules are needed to stabilise the structure (Figure 5.35c).

The connection rules described above are presented schematically in Figure 5.35, where two regular modules are used to demonstrate a principle for creating multi-module systems.

A general rule for creation of any tensegrity-based metamaterial or metastructure is illustrated in Figure 5.36. There are three key elements: a unit cell—usually a basic tensegrity module, a supercell—a repetitive cell consisting of several tensegrity modules in various arrangements, and a lattice—metamaterial or metastructure consisting of a number of supercells.

Within the supercells, single tensegrity modules can be arranged in different patterns to form lattices with certain properties. Depending on the module type, its geometry and orientation, and the way in which the modules are connected, lattices with various mechanical characteristics can be

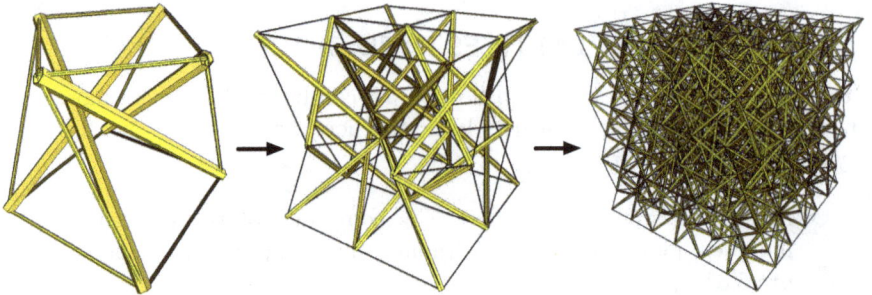

Figure 5.36 Creation of a 3D tensegrity lattice.

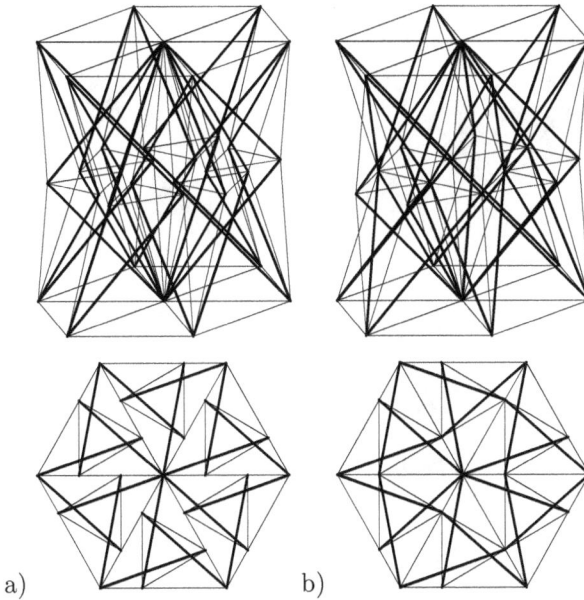

a)   b)

Figure 5.37 Three-strut simplex lattices: a) modules oriented consistently; b) modules oriented alternately.

obtained. Figure 5.37 shows two variants of the three-strut simplex lattice. In Figure 5.37a all modules are oriented consistently in accordance with their infinitesimal mechanisms, in Figure 5.37b the modules rotated clockwise and counterclockwise are arranged alternately to form a symmetrical system.

Further in this section, a study on tensegrity lattices based on the four-strut simplex module in various arrangements is presented. Similar analyses can be performed for other previously discussed modules. However, the aim of this section is to show that it is possible to obtain tensegrity lattices with extremal mechanical properties, and that these properties depend on various parameters, not to analyse all possible configurations of cellular structures based on tensegrity modules. All presented analyses are performed similarly to the analyses of single modules (see Section 5.2) and are based on the continuum model described in Section 4.2.

5.3.1 CONSTRUCTION OF MODULAR TENSEGRITY LATTICES

Connection of four-strut simplex modules leads to a modular system, where no extra cables are needed. However, depending on the geometry (variant of the module used - see Figure 5.12) and the orientation of particular modules, different extremal properties can be obtained. It is proved below that if all modules are arranged in accordance with their infinitesimal mechanisms, they form a structure (lattice) that has identical properties as its unit cells (single

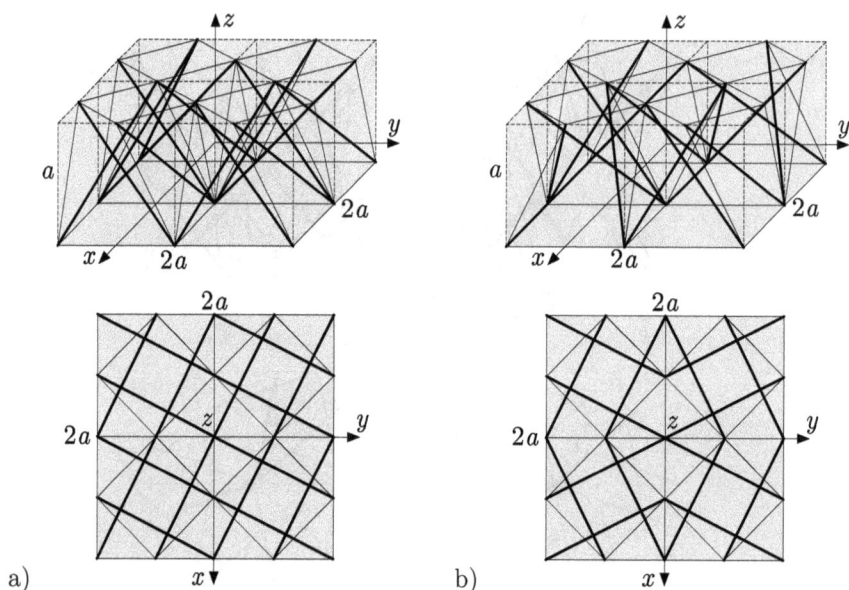

Figure 5.38 Configurations of the four-unit supercell: a) A—anisotropic system; b) B—orthotropic system.

modules) analysed separately. If, however, there is no kinematic compatibility between infinitesimal mechanisms of the unit cells, properties of the structure become different from the features of the single module.

Let us consider two configurations of the system consisting of four unit cells—four-strut simplex modules—connected via common cables of their lower bases and common nodes of their upper bases, further referred to as a four-unit supercell:

- configuration A—an anisotropic layout with four modules rotated clockwise (Figure 5.38a),
- configuration B—an orthotropic layout with two modules rotated clockwise and two counterclockwise to obtain symmetry (Figure 5.38b).

The supercells are constructed from the four-strut simplex modules inscribed into a cube of edge length a and therefore, the volume of the supercell in the formula for the strain energy of a solid determined according to the symmetric linear 3D elasticity theory (LTE) (see Section 4.2) should be taken as $4a^3$.

The elasticity matrix obtained from the continuum model for the anisotropic configuration A has a form

$$
\mathbf{E_A} = \begin{bmatrix}
e_{11} & e_{12} & e_{13} & e_{14} & 0 & 0 \\
& e_{11} & e_{13} & -e_{14} & 0 & 0 \\
& & e_{33} & 0 & 0 & 0 \\
& & & e_{12} & 0 & 0 \\
& & & & e_{13} & 0 \\
\text{sym.} & & & & & e_{13}
\end{bmatrix}, \tag{5.22}
$$

$$
e_{11} = \frac{2EA}{a^2}(0.314815 + 1.39827 \cdot k - 0.0794978 \cdot \sigma),
$$

$$
e_{12} = \frac{EA}{a^2}(0.296296 + 0.707107 \cdot k - 0.0134742 \cdot \sigma),
$$

$$
e_{13} = \frac{EA}{a^2}(0.740741 + 0.357771 \cdot k + 0.17247 \cdot \sigma), \tag{5.23}
$$

$$
e_{14} = \frac{EA}{a^2}(-0.222222 - 0.0808452 \cdot \sigma),
$$

$$
e_{33} = \frac{2EA}{a^2}(0.592593 + 1.43108 \cdot k - 0.17247 \cdot \sigma).
$$

The same matrix determined for the orthotropic configuration B takes a form

$$
\mathbf{E_B} = \begin{bmatrix}
e_{11} & e_{12} & e_{13} & 0 & 0 & 0 \\
& e_{11} & e_{13} & 0 & 0 & 0 \\
& & e_{33} & 0 & 0 & 0 \\
& & & e_{12} & 0 & 0 \\
& & & & e_{13} & 0 \\
\text{sym.} & & & & & e_{13}
\end{bmatrix}, \tag{5.24}
$$

$$
e_{11} = \frac{2EA}{a^2}(0.314815 + 1.39827 \cdot k - 0.0794978 \cdot \sigma),
$$

$$
e_{12} = \frac{EA}{a^2}(0.296296 + 0.707107 \cdot k - 0.0134742 \cdot \sigma),
$$

$$
e_{13} = \frac{EA}{a^2}(0.740741 + 0.357771 \cdot k + 0.17247 \cdot \sigma), \tag{5.25}
$$

$$
e_{33} = \frac{2EA}{a^2}(0.592593 + 1.43108 \cdot k - 0.17247 \cdot \sigma).
$$

It can be noticed that matrix $\mathbf{E_A}$ (Eq. 5.22) is identical to the elastic matrix obtained for the four-strut simplex module (see Section 5.2.2). Matrix $\mathbf{E_B}$ (Eq. 5.24), on the other hand, differs from the elasticity matrix of the unit cell. As the analysis of extremal mechanical properties is based on elasticity matrices, it can be concluded that depending on the arrangement of single

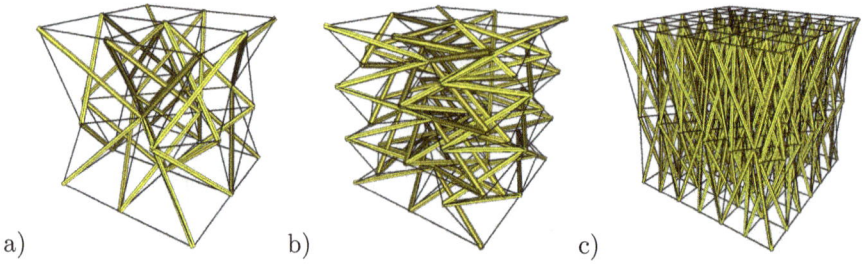

Figure 5.39 Four-strut simplex lattices: a) regular; b) very low; c) very high.

cells in the structure, properties of the lattice remain the same or become different from the features of the single module. Following this reasoning, the next section is focussed on the analysis of extremal properties of various tensegrity lattices, which are based on the four-strut simplex module in three geometrical variants. The anisotropic lattices are not considered there, as their extremal properties are the same as the properties of the four-strut simplex discussed in Section 5.2.2.

5.3.2   FOUR-STRUT SIMPLEX LATTICES

Let us consider three tensegrity lattices, inscribed into a cube $2a \times 2a \times 2a$:

- regular (R)—a lattice constructed from 8 regular four-strut simplex modules in an orthotropic layout (Figure 5.39a),
- very low (L)—a lattice constructed from 24 very low four-strut simplex modules in an orthotropic layout (Figure 5.39b),
- very high (H)—a lattice constructed from 72 very high four-strut simplex modules in an orthotropic layout (Figure 5.39c),

The regular lattice is constructed from two four-unit supercells in an orthotropic layout (see Figure 5.38b), where the upper layer of the system is created by putting the four-unit supercell upside-down and connecting it with the bottom layer via common cables (Figure 5.40). Very low and very high lattices are constructed similarly, but more modules are connected to fill the cube $2a \times 2a \times 2a$.

The elasticity matrices obtained from the continuum analysis of the considered lattices have the same orthotropic structure, but different coefficients:

$$\mathbf{E}_{\text{R/L/H}} = \begin{bmatrix} e_{11} & e_{12} & e_{13} & 0 & 0 & 0 \\ & e_{11} & e_{13} & 0 & 0 & 0 \\ & & e_{33} & 0 & 0 & 0 \\ & & & e_{12} & 0 & 0 \\ & & & & e_{13} & 0 \\ \text{sym.} & & & & & e_{13} \end{bmatrix}, \qquad (5.26)$$

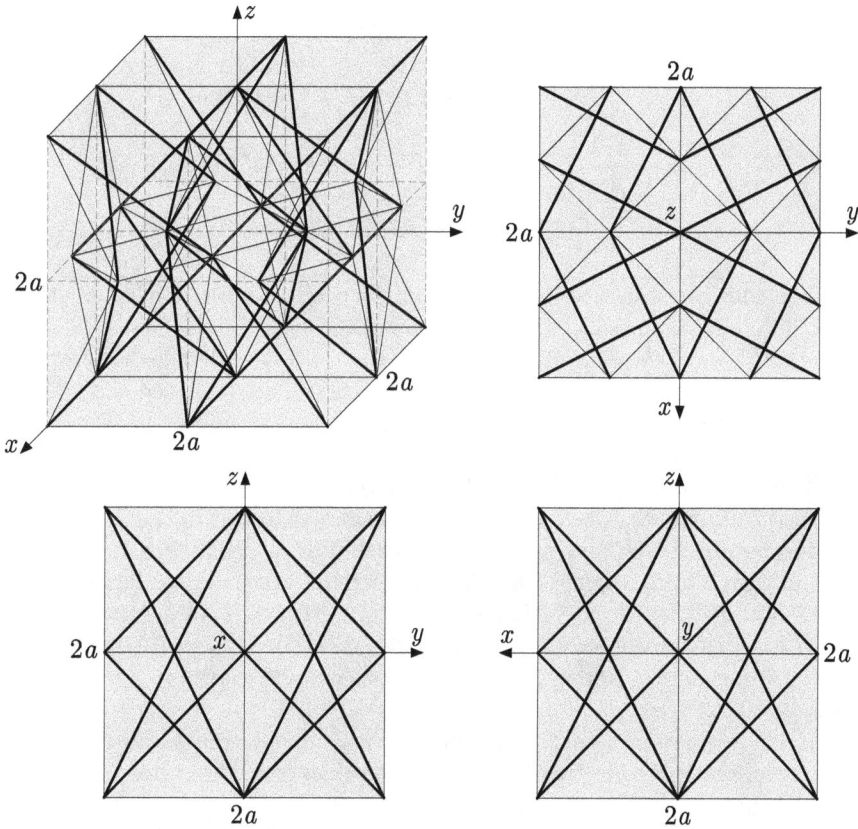

Figure 5.40 Geometry of the regular four-strut simplex lattice.

$$e_{R,11} = \frac{2EA}{a^2}(0.314815 + 1.39827 \cdot k - 0.0794978 \cdot \sigma),$$

$$e_{R,12} = \frac{EA}{a^2}(0.296296 + 0.707107 \cdot k - 0.0134742 \cdot \sigma),$$

$$e_{R,13} = \frac{EA}{a^2}(0.740741 + 0.357771 \cdot k + 0.17247 \cdot \sigma),$$

$$e_{R,33} = \frac{2EA}{a^2}(0.592593 + 1.43108 \cdot k - 0.17247 \cdot \sigma), \qquad (5.27)$$

$$e_{L,11} = \frac{2EA}{a^2}(2.00729 + 4.92471 \cdot k - 0.135911 \cdot \sigma),$$

$$e_{L,12} = \frac{EA}{a^2}(1.88921 + 2.12132 \cdot k + 0.223898 \cdot \sigma),$$

$$e_{L,13} = \frac{EA}{a^2}(0.524781 + 0.76805 \cdot k + 0.0479246 \cdot \sigma),$$

$$e_{L,33} = \frac{2EA}{a^2}(0.0466472 + 0.34135 \cdot k - 0.0479246 \cdot \sigma), \qquad (5.28)$$

$$e_{H,11} = \frac{2EA}{a^2}(0.0971325 + 4.06732 \cdot k - 0.217699 \cdot \sigma),$$

$$e_{H,12} = \frac{EA}{a^2}(0.0914188 + 2.12132 \cdot k - 0.134146 \cdot \sigma),$$

$$e_{H,13} = \frac{EA}{a^2}(2.05692 + 0.47987 \cdot k + 0.569545 \cdot \sigma),$$

$$e_{H,33} = \frac{2EA}{a^2}(14.8098 + 17.27522 \cdot k - 0.569545 \cdot \sigma). \tag{5.29}$$

The considered lattices, although inscribed into a cube of edge length $2a$, have different volumes of structural members: $V_R = aA_{strut}(48.00 + 62.15 \cdot k)$, $V_L = aA_{strut}(112.00 + 125.21 \cdot k)$, $V_H = aA_{strut}(307.35 + 260.26 \cdot k)$. Extremal properties of the systems, discussed in this section, are analysed according to the eigenvalues and eigenvectors of elasticity matrices under the assumption of an equal total volume $V = V_R$.

The eigenvalues $\lambda_i > 0$ ($i = 1, 2, \ldots, 6$) and orthogonal eigenvectors $\mathbf{w}_i$ of elasticity matrices obtained for the considered tensegrity lattices depend on the parameters k and σ. They are presented in Figures 5.41 and 5.42, and in Table 5.12. Figure 5.41 depicts lines of extremal mechanical properties obtained for three variants of the four-strut simplex lattice and Figure 5.42—distribution of eigenvalues determined for $k_{extr} = 0.1$ and the following values of σ_{extr} depending on the considered variant: $\sigma_{R,extr} = 0.5464$, $\sigma_{L,extr} = 0.5268$, $\sigma_{H,extr} = 1.3734$.

It can be noticed that although the considered configurations are orthotropic, as opposed to their unit cells, the values of σ_{extr} obtained for lattices are the same as the corresponding values determined for tensegrity modules. Their extremal mechanical properties, however, differ. The regular and very low four-strut simplex lattices are unimode—there is only one eigenvalue that is equal to zero, and others are significantly bigger (as opposed to their unit cells, which are quasi bimode and bimode), and the very high variant can be

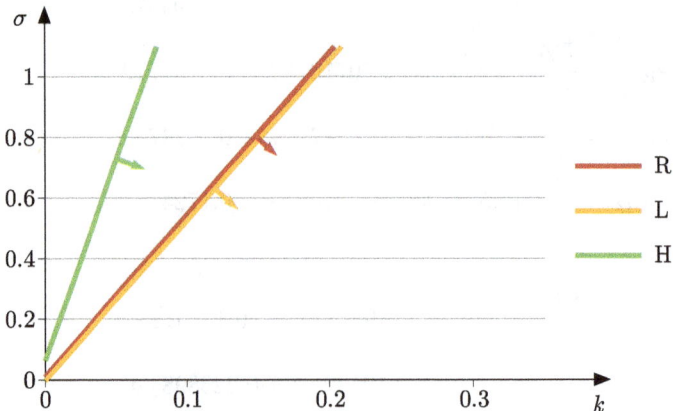

Figure 5.41 Lines of extremal properties of the regular, very low and very high four-strut simplex lattice.

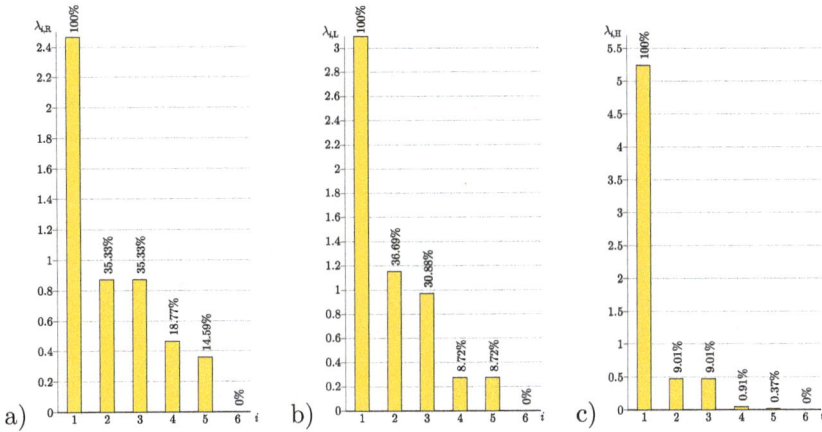

Figure 5.42 Distribution of eigenvalues for three variants of the four-strut simplex lattice: a) regular; b) very low; c) very high (multiplier EA/a^2).

defined as trimode—one eigenvalue is equal to zero and two others are smaller than 1% of the maximum eigenvalue obtained for this variant (the unit cell is quasi trimode). Moreover, the high lattice exhibits the biggest differences between the eigenvalue corresponding to the stiff mode of deformation and the ones indicating medium modes (eigenvalues around 9% of $\lambda_{\max,\mathrm{H}}$).

The soft mode of deformation of the regular four-strut simplex lattice (Figure 5.43a), represented by the eigenvector $\mathbf{w}_{6,\mathrm{R}}$, is volumetric with various signs, and the extension in X_3 direction exceeds by 32% the contraction in other directions. The stiff mode (Figure 5.44a), represented by the eigenvector $\mathbf{w}_{1,\mathrm{R}}$, is volumetric with a uniform sign, and the extension in X_3 direction is 54% bigger than in others. It can be noticed that although the single regular four-strut simplex module is anisotropic and the analysed regular lattice is orthotropic, the soft and stiff modes of deformation obtained for both systems are very similar, both qualitatively and quantitatively.

Other variants of the four-strut simplex lattice exhibit different extremal behaviours. The soft mode of deformation of the low system (Figure 5.43b), represented by the eigenvector $\mathbf{w}_{6,\mathrm{L}}$, is also volumetric with various signs, but the extension in X_3 direction dominates over two other directions. The stiff

Table 5.12

Extremal properties of three four-strut simplex lattices.

Module	σ_{extr}	Extremal Properties
regular	0.5464	unimode
very low	0.5268	unimode
very high	1.3734	trimode

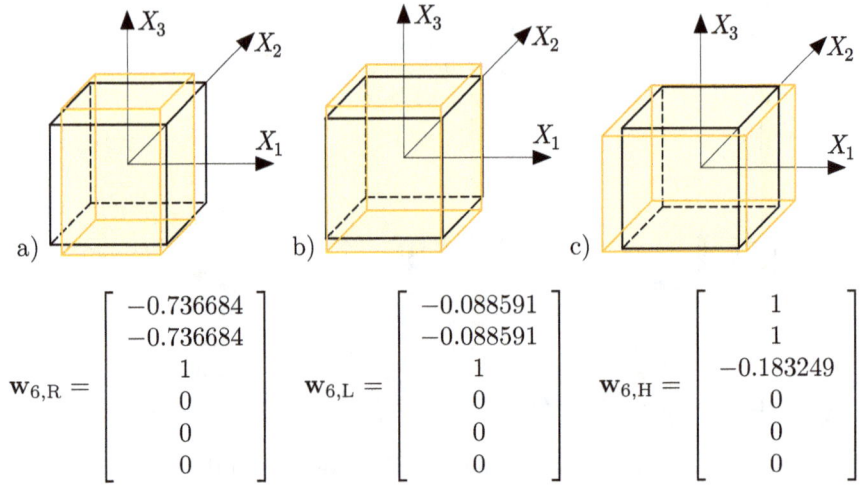

$$\mathbf{w}_{6,\mathrm{R}} = \begin{bmatrix} -0.736684 \\ -0.736684 \\ 1 \\ 0 \\ 0 \\ 0 \end{bmatrix} \qquad \mathbf{w}_{6,\mathrm{L}} = \begin{bmatrix} -0.088591 \\ -0.088591 \\ 1 \\ 0 \\ 0 \\ 0 \end{bmatrix} \qquad \mathbf{w}_{6,\mathrm{H}} = \begin{bmatrix} 1 \\ 1 \\ -0.183249 \\ 0 \\ 0 \\ 0 \end{bmatrix}$$

Figure 5.43 Soft modes of deformation of three variants of the four-strut simplex lattice: a) regular; b) very low; c) very high.

mode (Figure 5.44b), represented by the eigenvector $\mathbf{w}_{1,\mathrm{L}}$, is volumetric with a uniform sign and the extensions in X_1 and X_2 directions are dominant.

The soft mode of deformation of the high system (Figure 5.43c), represented by the eigenvector $\mathbf{w}_{6,\mathrm{H}}$, is volumetric with various signs, and the extensions in X_1 and X_2 directions dominate over the contraction in X_3 direction. The stiff mode (Figure 5.44c), represented by the eigenvector $\mathbf{w}_{1,\mathrm{H}}$, is volumetric with a uniform sign and the extension in X_3 direction is dominant.

Apart from the analysis of eigenvalues and deformation modes, a continuum approach described in Section 4.2 allows us to determine technical coefficients of the systems. Below, example of such an analysis based on the regular four-strut simplex lattice is presented. First, in order to find mechanical characteristics of the structure, an inverse matrix $\mathbf{H} = \mathbf{E}^{-1}$ needs to be determined. The matrix has seven independent coefficients:

$$\mathbf{H}_{\mathrm{R}} = \mathbf{E}_{\mathrm{R}}^{-1} = \begin{bmatrix} \dfrac{1}{E_1} & -\dfrac{\nu_{12}}{E_1} & -\dfrac{\nu_{31}}{E_3} & 0 & 0 & 0 \\[2mm] -\dfrac{\nu_{12}}{E_1} & \dfrac{1}{E_1} & -\dfrac{\nu_{31}}{E_3} & 0 & 0 & 0 \\[2mm] -\dfrac{\nu_{13}}{E_1} & -\dfrac{\nu_{13}}{E_1} & \dfrac{1}{E_3} & 0 & 0 & 0 \\[2mm] 0 & 0 & 0 & \dfrac{1}{G_1} & 0 & 0 \\[2mm] 0 & 0 & 0 & 0 & \dfrac{1}{G_2} & 0 \\[2mm] 0 & 0 & 0 & 0 & 0 & \dfrac{1}{G_2} \end{bmatrix}, \qquad (5.30)$$

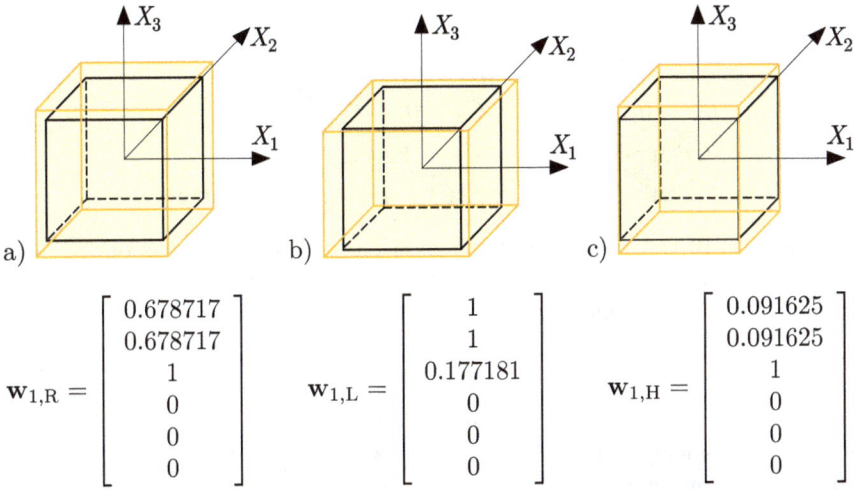

Figure 5.44 Stiff modes of deformation of three variants of the four-strut simplex lattice: a) regular; b) very low; c) very high.

with the following values:

$$E_1 = E_2 = \frac{(e_{11} - e_{12})(2e_{13}^2 - (e_{11} + e_{12})e_{33})}{e_{13}^2 - e_{11}e_{33}},$$

$$E_3 = \frac{-2e_{13}^2 + (e_{11} + e_{12})e_{33}}{e_{11} + e_{12}},$$

$$G_1 = e_{12},$$

$$G_2 = G_3 = e_{13},$$

$$\nu_{12} = \nu_{21} = \frac{e_{13}^2 - e_{12}e_{33}}{e_{13}^2 - e_{11}e_{33}},$$

$$\nu_{13} = \nu_{23} = \frac{-e_{13}(e_{11} - e_{12})}{e_{13}^2 - e_{11}e_{33}},$$

$$\nu_{31} = \nu_{32} = \frac{e_{13}}{e_{11} + e_{12}}.$$

Symmetry of the matrix $\mathbf{H_R}$ leads to the condition

$$\frac{\nu_{13}}{E_1} = \frac{\nu_{31}}{E_3}. \tag{5.31}$$

Moreover, due to the fact that the matrices $\mathbf{E_R}$ and $\mathbf{H_R}$ are positive definite, the following conditions limiting the values of technical coefficients are obtained:

$$E_1 > 0, \quad E_2 > 0, \quad E_3 > 0,$$
$$G_1 > 0, \quad G_2 > 0, \quad G_3 > 0,$$
$$\nu_{12}\nu_{21} < 1, \quad \nu_{13}\nu_{31} < 1, \quad \nu_{23}\nu_{32} < 1, \tag{5.32}$$
$$\nu_{12}\nu_{21} + \nu_{13}\nu_{31} + \nu_{23}\nu_{32} + \nu_{12}\nu_{31}\nu_{23} + \nu_{21}\nu_{13}\nu_{32} < 1.$$

Analysis of the above limiting conditions and the domains of the determined technical coefficients leads to the general condition that is a common domain for all mechanical characteristics:

$$\sigma < 5.19956k. \tag{5.33}$$

Using the results obtained from the presented continuum analysis, an influence of element properties and self-stress on mechanical characteristics of the regular four-strut simplex lattice can be determined. Figures 5.45–5.50 show how the selected technical coefficients depend on the defined parameters k and σ. It can be noticed that all analysed coefficients are prone to structural control—their values may be controlled by adjusting either the properties of struts and cables or the values of prestressing forces in structural members. This feature distinguishes smart structural systems from the traditional ones. While a typical structure exhibits certain properties which are constant (assuming that the rheological phenomena are neglected), the characteristics of a smart structure can be adjusted in an operating system.

Due to the fact that parameter k depends on physical and geometrical properties of cables and struts, it is usually fixed for the whole structure at a certain level. The role of self-stress parameter σ is different, as it can be adjusted during exploitation of the system to control the values of elastic coefficients. It can be noticed in Figures 5.45 and 5.46 that the influence of both parameters k and σ on the Young's and shear modulus of the lattice is significant.

One of the features that characterise extremal systems are negative values of Poisson's ratio. In the case of the proposed four-strut simplex lattice, Poisson's ratios $\nu_{12} = \nu_{21}$ can not only have negative values but they can also change sign, what is demonstrated in Figures 5.47 and 5.48. It means that the proposed structure behaves differently depending on the adopted parameters k and σ. This is a unique feature of smart systems—they can act like standard structures with positive Poisson's ratios and, at the same time, can be changed into extremal systems with negative values of these mechanical characteristics.

Other Poisson's ratios are always positive (Figures 5.49 and 5.50) but sensitive to k and σ.

Tensegrity systems described above can be used to build modular lattices of any volume. Properties of such structures will be the same as the properties of the analysed lattices inscribed into a cube $2a \times 2a \times 2a$.

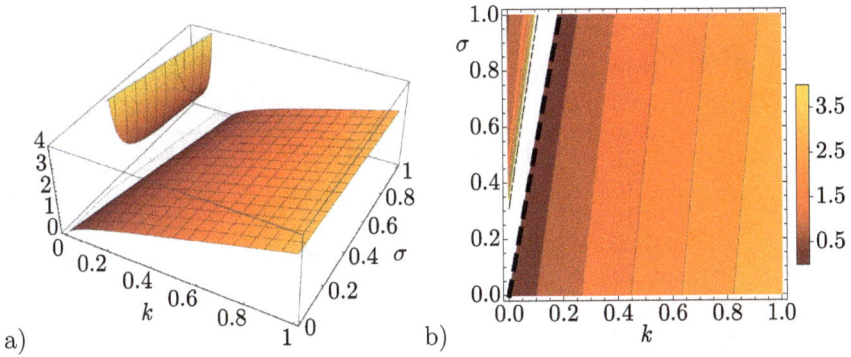

Figure 5.45 Young's modulus E_1 (divided by the factor EA/a^2) with the limiting condition $E_1 > 0$: a) 3D plot; b) contour plot.

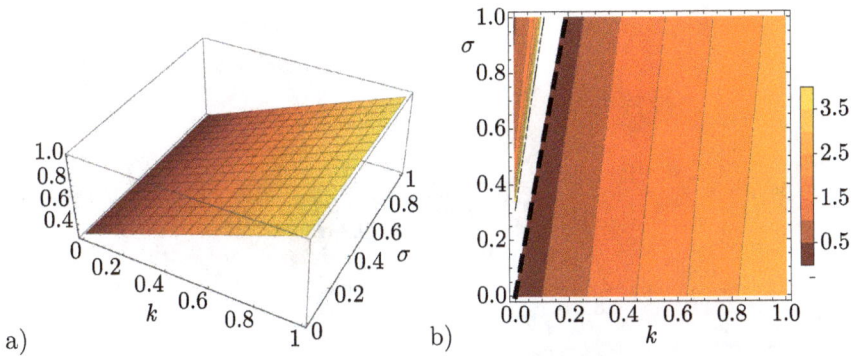

Figure 5.46 Shear modulus G_1 (divided by the factor EA/a^2) with the limiting condition $G_1 > 0$: a) 3D plot; b) contour plot.

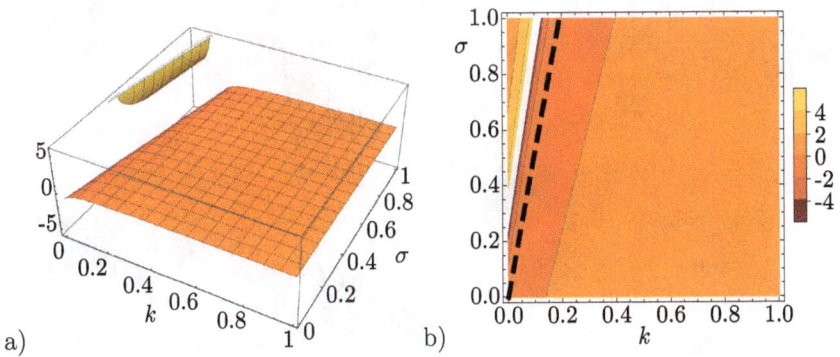

Figure 5.47 Poisson's ratio $\nu_{12} = \nu_{21}$: a) 3D plot; b) contour plot.

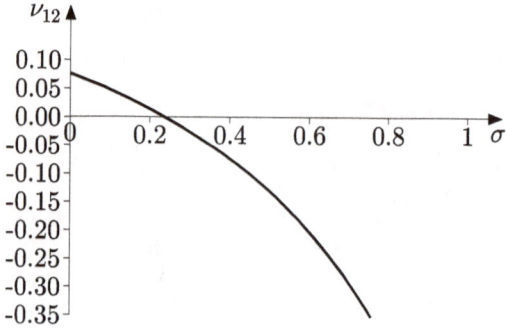

Figure 5.48 Poisson's ratio $\nu_{12} = \nu_{21}$ changing sign (plot for $k = 0.2$).

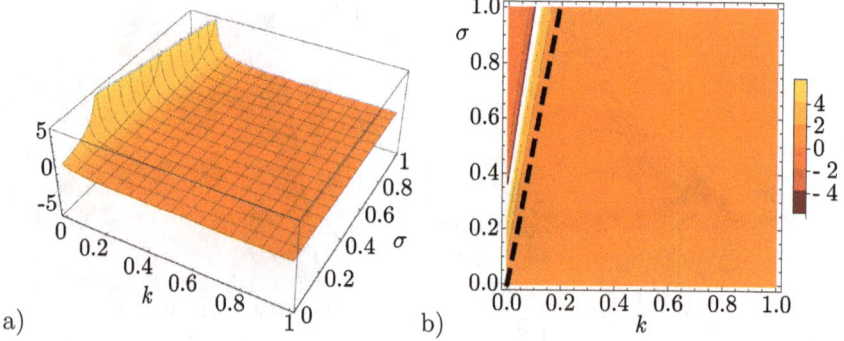

Figure 5.49 Poisson's ratio $\nu_{13} = \nu_{23}$: a) 3D plot; b) contour plot.

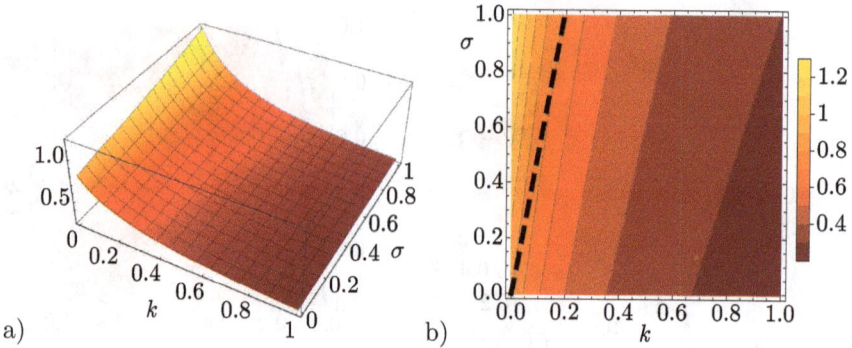

Figure 5.50 Poisson's ratio $\nu_{31} = \nu_{32}$: a) 3D plot; b) contour plot.

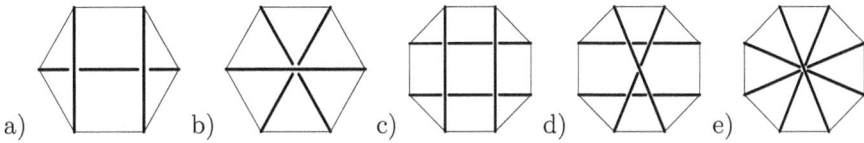

Figure 5.51 2D tensegrity modules: a) hexagon 1; b) hexagon 2; c) octagon 1; d) octagon 2; e) octagon 3.

Examples of analyses presented in Sections 5.3.1–5.3.2 are very important from the practical point of view. It was shown that basic tensegrity modules considered in Section 5.2 can be used to build more complex systems, such as modular metamaterials based on supercells. Similarly to their unit cells, such systems exhibit extremal mechanical behaviour, and can therefore be applied as high-performance metamaterials in many engineering solutions. It is worth noticing, that the study performed for tensegrity lattices was extended compared to tensegrity modules—it included identification of technical coefficients of the systems, whose analysis is essential from the engineering point of view. These characteristics allow us to understand behaviour of the systems in real applications and therefore, are commonly used in engineering practice.

5.4 TENSEGRITY MODULES IN 2D SPACE

This section is focussed on the analysis of extremal behaviour of basic 2D tensegrity modules. Such systems may be less practical from the engineering point of view (especially considering bigger scales). However, the identification of extremal properties of 2D modules leads to closed-form solutions for eigenvalues that describe these properties, and therefore, it adds value to the understanding of the whole concept of extremal materials.

Below, a study on extremal mechanical properties of five 2D tensegrity modules is presented: two hexagonal and three octagonal modules (Figure 5.51).

The analyses are based on the control of two parameters k and σ, which were introduced in Section 5.1.

In order to analyse extremal properties of the structures, the following features are described for 2D modules:

- equivalent elasticity matrix,
- eigenvalues and corresponding eigenvectors,
- distribution of eigenvalues depending on parameters k and σ,
- deformation modes.

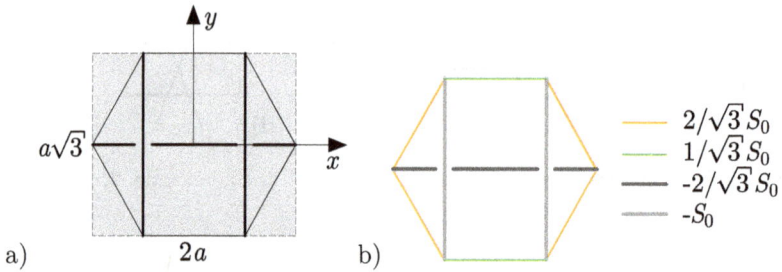

Figure 5.52 Hexagon 1: a) module inscribed into a rectangle; b) self-stress state.

5.4.1 HEXAGON 1

The analysed 2D hexagonal module (H1) is presented in Figure 5.52a, with its geometry described in Table 5.13.

The module has one infinitesimal mechanism and one corresponding self-stress state (Figure 5.52b)—self-stress is expressed by relative forces in struts and cables with a multiplier S_0.

The elasticity matrix obtained from the continuum model has a form

$$\mathbf{E}_{H1} = \frac{EA}{ah}\frac{1}{8\sqrt{3}}\begin{bmatrix} 8+9k & 3k & 0 \\ 3k & 8\sqrt{3}+9k & 0 \\ 0 & 0 & 3k \end{bmatrix} + \frac{S_0}{ah}\frac{1}{4}\begin{bmatrix} 1 & -1 & 0 \\ -1 & 1 & 0 \\ 0 & 0 & -1 \end{bmatrix}.$$

$$(5.34)$$

Spectral analysis of matrix $\mathbf{E}$ allows us to find eigenvalues, which describe extremal properties of the module:

Table 5.13

Geometry of hexagon 1.

Node No.	x	y	Scheme
1	$-a$	0	
2	$-a/2$	$a\sqrt{3}/2$	
3	$a/2$	$a\sqrt{3}/2$	
4	a	0	
5	$a/2$	$-a\sqrt{3}/2$	
6	$-a/2$	$-a\sqrt{3}/2$	

$$\lambda_1 = \frac{EA}{24ah} \left(\begin{array}{c} 12 + 4\sqrt{3} + 9\sqrt{3}k + 6\sigma+ \\ \hline + \sqrt{96\left(2-\sqrt{3}\right) + 27k^2 - 36\sqrt{3}k\sigma + 36\sigma^2} \end{array} \right) > 0,$$

$$\lambda_2 = \frac{EA}{24ah} \left(\begin{array}{c} 12 + 4\sqrt{3} + 9\sqrt{3}k + 6\sigma+ \\ \hline - \sqrt{96\left(2-\sqrt{3}\right) + 27k^2 - 36\sqrt{3}k\sigma + 36\sigma^2} \end{array} \right) > 0, \qquad (5.35)$$

$$\lambda_3 = \frac{EA}{8ah} \left(\sqrt{3}k - 2\sigma \right).$$

Figure 5.53 shows how the parameters k and σ influence the eigenvalues determined above. It can be noticed that within the analysed range of parameters $k \in (0,1)$ and $\sigma \in (0,1)$, only the third eigenvalue can reach zero, the other two are always positive. The plot of λ_3 (Figure 5.53c) is limited to positive values of the third eigenvalue, in order to demonstrate clearly the line of extremal properties, where $\lambda_3 = 0$.

The module can be identified as unimode with one soft mode of deformation obtained for

$$\sigma = \frac{k\sqrt{3}}{2}. \qquad (5.36)$$

After substitution of Eq. 5.36 into Eq. 5.35, the following eigenvalues are obtained with corresponding eigenvectors shown in Figure 5.54:

$$\lambda_1 = \frac{EA}{2ah}(\sqrt{3}k + 2), \quad \lambda_2 = \frac{EA}{2ah\sqrt{3}}(3k + 2), \quad \lambda_3 = 0. \qquad (5.37)$$

The stiff mode (Figure 5.54a), represented by the eigenvector $\mathbf{w}_{1,\mathrm{H1}}$, is volumetric with an extension in X_2 direction. Another stiff mode (Figure 5.54b), represented by the eigenvector $\mathbf{w}_{2,\mathrm{H1}}$, is also volumetric with an extension in X_1 direction. The soft mode of deformation (Figure 5.54c) is represented by the eigenvector $\mathbf{w}_{3,\mathrm{H1}}$ and it is a shear deformation in X_1–X_2 plane.

5.4.2 HEXAGON 2

The analysed 2D hexagonal module (H2) is presented in Figure 5.55a, with its geometry described in Table 5.14.

The module has one infinitesimal mechanism and one corresponding self-stress state (Figure 5.55b)—self-stress is expressed by relative forces in struts and cables with a multiplier S_0.

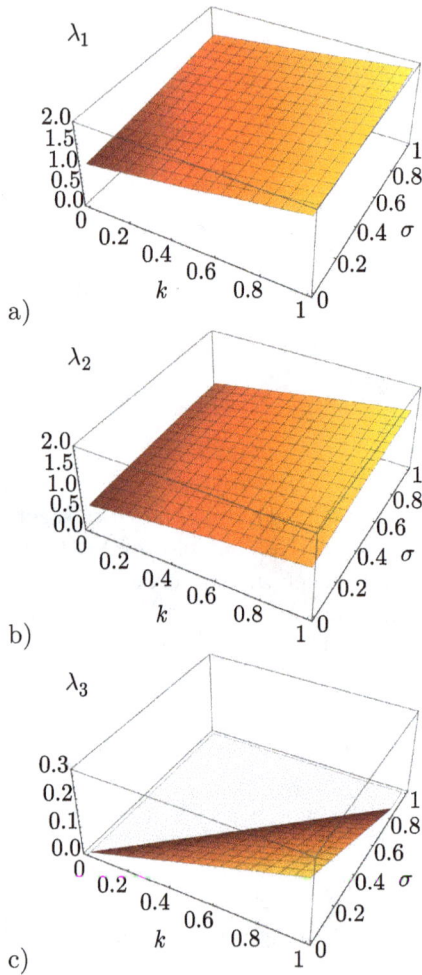

Figure 5.53 Distribution of eigenvalues for hexagon 1: a) λ_1; b) λ_2; c) λ_3.

The elasticity matrix obtained from the continuum model has a form

$$\mathbf{E}_{\mathrm{H2}} = \frac{EA}{ah} \frac{3(1+k)}{8\sqrt{3}} \begin{bmatrix} 3 & 1 & 0 \\ 1 & 3 & 0 \\ 0 & 0 & 1 \end{bmatrix}. \tag{5.38}$$

$$\mathbf{w}_{1,\mathrm{H1}} = \begin{bmatrix} 0 \\ 1 \\ 0 \end{bmatrix} \qquad \mathbf{w}_{2,\mathrm{H1}} = \begin{bmatrix} 1 \\ 0 \\ 0 \end{bmatrix} \qquad \mathbf{w}_{3,\mathrm{H1}} = \begin{bmatrix} 0 \\ 0 \\ 1 \end{bmatrix}$$

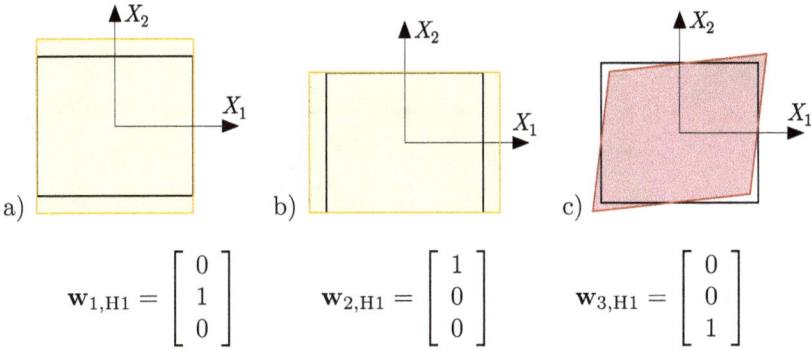

Figure 5.54 Deformation modes of hexagon 1: a) stiff mode corresponding to λ_1; b) stiff mode corresponding to λ_2; c) soft mode corresponding to λ_3.

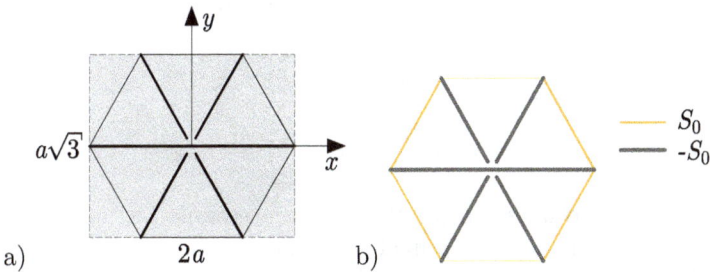

Figure 5.55 Hexagon 2: a) module inscribed into a rectangle; b) self-stress state.

Table 5.14

Geometry of hexagon 2.

Node No.	x	y	Scheme
1	$-a$	0	
2	$-a/2$	$a\sqrt{3}/2$	
3	$a/2$	$a\sqrt{3}/2$	
4	a	0	
5	$a/2$	$-a\sqrt{3}/2$	
6	$-a/2$	$-a\sqrt{3}/2$	

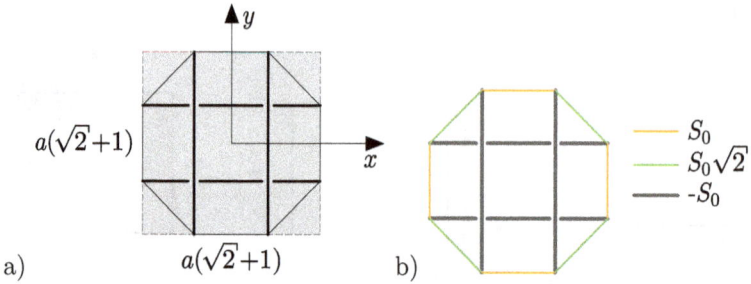

Figure 5.56 Octagon 1: a) module inscribed into a square; b) self-stress state.

It can be noticed that contrary to hexagon 1, matrix $\mathbf{E}$ of hexagon 2 does not depend on self-stress forces. Spectral analysis of matrix $\mathbf{E}$ allows us to find the following eigenvalues:

$$
\begin{aligned}
\lambda_1 &= \frac{EA\sqrt{3}}{2ah}(1+k) > 0, \\
\lambda_2 &= \frac{EA\sqrt{3}}{4ah}(1+k) > 0, \\
\lambda_3 &= \frac{EA\sqrt{3}}{8ah}(1+k) > 0.
\end{aligned}
\tag{5.39}
$$

All three eigenvalues are positive and the module has no soft modes of deformation—it should be classified as a nullmode material.

5.4.3 OCTAGON 1

The analysed 2D octagonal module (O1) is presented in Figure 5.56a, with its geometry described in Table 5.15.

The module has two infinitesimal mechanisms and one self-stress state (Figure 5.56b)—self-stress is expressed by relative forces in struts and cables with a multiplier S_0.

The elasticity matrix obtained from the continuum model has a form

$$
\mathbf{E}_{O1} = \frac{EA}{ah}
\begin{bmatrix}
2\left(\sqrt{2}-1\right)+\left(9-6\sqrt{2}\right)k & \left(3-2\sqrt{2}\right)k & 0 \\
\left(3-2\sqrt{2}\right)k & 2\left(\sqrt{2}-1\right)+\left(9-6\sqrt{2}\right)k & 0 \\
0 & 0 & \left(3-2\sqrt{2}\right)k
\end{bmatrix} + \\
+ \frac{S_0}{ah}\left(3\sqrt{2}-4\right)
\begin{bmatrix}
1 & -1 & 0 \\
-1 & 1 & 0 \\
0 & 0 & -1
\end{bmatrix}.
$$

$$
\tag{5.40}
$$

Spectral analysis of matrix $\mathbf{E}$ allows us to find eigenvalues and corresponding eigenvectors, which describe extremal properties of the module:

Table 5.15

Geometry of octagon 1.

Node No.	x	y	Scheme
1	$-a(1+\sqrt{2})/2$	$-a/2$	
2	$-a(1+\sqrt{2})/2$	$a/2$	
3	$-a/2$	$a(1+\sqrt{2})/2$	
4	$a/2$	$a(1+\sqrt{2})/2$	
5	$a(1+\sqrt{2})/2$	$a/2$	
6	$a(1+\sqrt{2})/2$	$-a/2$	
7	$a/2$	$-a(1+\sqrt{2})/2$	
8	$-a/2$	$-a(1+\sqrt{2})/2$	

$$\lambda_1 = \frac{EA}{ah}\left[2\left(\sqrt{2}-1\right)+4\left(3-2\sqrt{2}\right)k\right] > 0,$$

$$\lambda_2 = \frac{EA}{ah}\frac{2}{2-\sqrt{2}}\left(5\sqrt{2}-7\right)\left(\sqrt{2}+2+\sqrt{2}k+2\sigma\right) > 0, \qquad (5.41)$$

$$\lambda_3 = \frac{EA}{ah}\left[\left(3-2\sqrt{2}\right)k+\left(4-3\sqrt{2}\right)\sigma\right].$$

Figure 5.57 shows how the parameters k and σ influence the eigenvalues determined above. The plot of λ_3 (Figure 5.57c) is limited to positive values of the third eigenvalue, in order to demonstrate clearly the line of extremal properties, where $\lambda_3 = 0$.

The module can be identified as unimode with one soft mode of deformation obtained for

$$\sigma = \frac{k\sqrt{2}}{2}. \qquad (5.42)$$

After substitution of Eq. 5.42 into Eq. 5.41, the following eigenvalues are obtained with corresponding eigenvectors shown in Figure 5.58:

$$\lambda_1 = \lambda_2 = \frac{EA}{ah}\left[2\left(\sqrt{2}-1\right)+4\left(3-2\sqrt{2}\right)k\right], \qquad \lambda_3 = 0. \qquad (5.43)$$

It is worth noticing that although the expressions for λ_1 and λ_2 (Eq. 5.41) were different, after substitution of Eq. 5.42 into Eq. 5.41, equal values of both eigenvalues corresponding to extremal behaviour of the module were obtained.

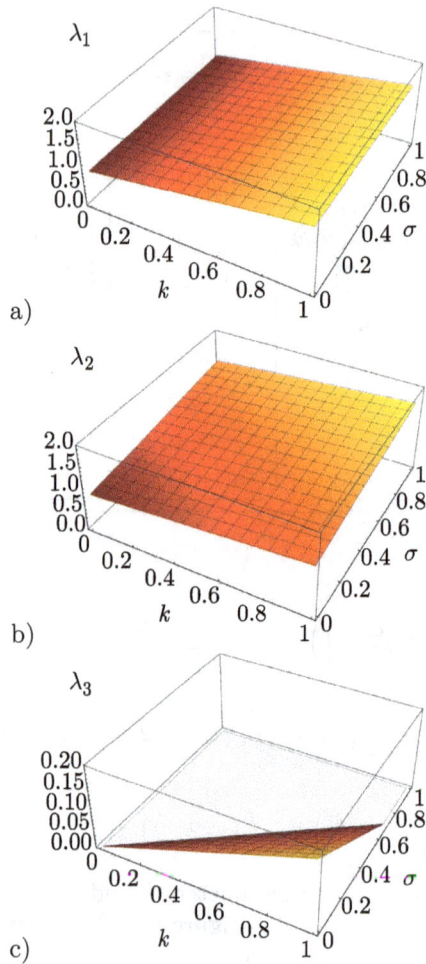

Figure 5.57 Distribution of eigenvalues for octagon 1: a) λ_1; b) λ_2; c) λ_3.

The stiff mode (Figure 5.58a), represented by the eigenvector $\mathbf{w}_{1,O1}$, is volumetric with an extension in X_1 direction. Another stiff mode (Figure 5.58b), represented by the eigenvector $\mathbf{w}_{2,O1}$, is also volumetric with an extension in X_2 direction. The soft mode of deformation (Figure 5.58c) is represented by the eigenvector $\mathbf{w}_{3,O1}$ and it is a shear deformation in X_1–X_2 plane.

5.4.4   OCTAGON 2

The analysed 2D octagonal module (O2) is presented in Figure 5.59a, with its geometry described in Table 5.16.

$$\mathbf{w}_{1,O1} = \begin{bmatrix} 1 \\ 0 \\ 0 \end{bmatrix} \qquad \mathbf{w}_{2,O1} = \begin{bmatrix} 0 \\ 1 \\ 0 \end{bmatrix} \qquad \mathbf{w}_{3,O1} = \begin{bmatrix} 0 \\ 0 \\ 1 \end{bmatrix}$$

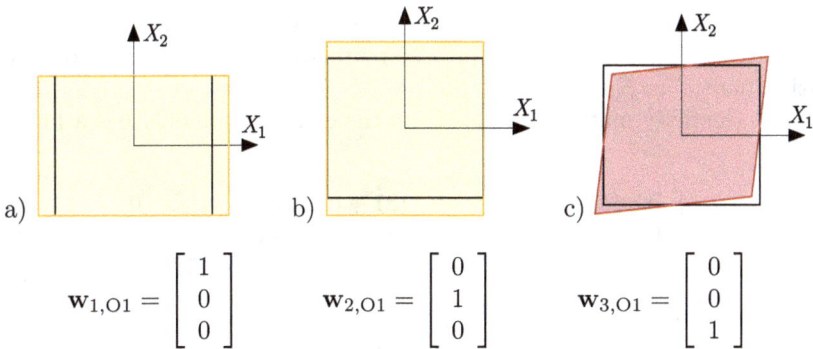

Figure 5.58 Deformation modes of octagon 1: a) stiff mode corresponding to λ_1; b) stiff mode corresponding to λ_2; c) soft mode corresponding to λ_3.

Table 5.16
Geometry of octagon 2.

Node No.	x	y	Scheme
1	$-a(1+\sqrt{2})/2$	$-a/2$	
2	$-a(1+\sqrt{2})/2$	$a/2$	
3	$-a/2$	$a(1+\sqrt{2})/2$	
4	$a/2$	$a(1+\sqrt{2})/2$	
5	$a(1+\sqrt{2})/2$	$a/2$	
6	$a(1+\sqrt{2})/2$	$-a/2$	
7	$a/2$	$-a(1+\sqrt{2})/2$	
8	$-a/2$	$-a(1+\sqrt{2})/2$	

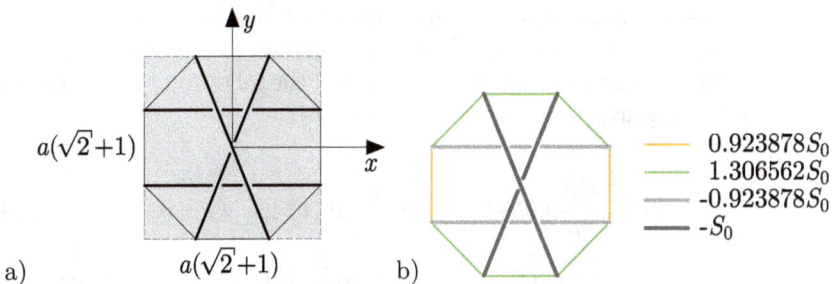

———	$0.923878S_0$
———	$1.306562S_0$
———	$-0.923878S_0$
———	$-S_0$

Figure 5.59 Octagon 2: a) module inscribed into a square; b) self-stress state.

The module has two infinitesimal mechanisms and one self-stress state (Figure 5.59b)—self-stress is expressed by relative forces in struts and cables with a multiplier S_0.

The elasticity matrix obtained from the continuum model has a form

$$
\mathbf{E}_{O2} = \frac{EA}{ah}
\begin{bmatrix}
0.8477 + 0.5147k & 0.1121 + 0.1716k & 0 \\
0.1121 + 0.1716k & 0.6533 + 0.5147k & 0 \\
0 & 0 & 0.1121 + 0.1716k
\end{bmatrix} +
$$

$$
+ 0.1121 \frac{S_0}{ah}
\begin{bmatrix}
1 & -1 & 0 \\
-1 & 1 & 0 \\
0 & 0 & -1
\end{bmatrix}.
$$

$$(5.44)$$

Spectral analysis of matrix $\mathbf{E}$ allows us to find eigenvalues and corresponding eigenvectors, which describe extremal properties of the module:

$$
\lambda_1 = \frac{0.1177EA}{ah}
\left(
\begin{array}{l}
6.3735 + 4.3713k + 0.9519\sigma + \\
+ \sqrt{1.5874 + 2.774k + 2.1232k^2 - 1.8122\sigma - 2.774k\sigma + 0.9061\sigma^2}
\end{array}
\right) > 0,
$$

$$
\lambda_2 = \frac{0.1177EA}{ah}
\left(
\begin{array}{l}
6.3735 + 4.3713k + 0.9519\sigma + \\
- \sqrt{1.5874 + 2.774k + 2.1232k^2 - 1.8122\sigma - 2.774k\sigma + 0.9061\sigma^2}
\end{array}
\right) > 0,
$$

$$
\lambda_3 = \frac{EA}{ah} (0.1121 + 0.1716k - 0.1121\sigma).
$$

$$(5.45)$$

Figure 5.60 shows how the parameters k and σ influence the eigenvalues determined above.

It can be noticed that within the analysed range of parameters $k \in (0, 1)$ and $\sigma \in (0, 1)$, no eigenvalue can reach zero. However, the third eigenvalue can reach zero in a particular case, for $\sigma = 1$ and $k = 0$. Such a scenario could not be realised in practice, as it would mean that the cables stiffness is zero. Nevertheless, in theory this module can be classified as a particular unimode material with one soft mode of deformation and the following eigenvalues with corresponding eigenvectors shown in Figure 5.61:

$$\lambda_1 = \frac{EA}{ah}0.9594, \quad \lambda_2 = \frac{EA}{ah}0.7654, \quad \lambda_3 = 0. \qquad (5.46)$$

The stiff mode (Figure 5.61a), represented by the eigenvector $\mathbf{w}_{1,O2}$, is volumetric with an extension in X_1 direction. Another stiff mode (Figure 5.61b), represented by the eigenvector $\mathbf{w}_{2,O2}$, is also volumetric with an extension in X_2 direction. The soft mode of deformation (Figure 5.61c) is represented by the eigenvector $\mathbf{w}_{3,O2}$ and it is a shear deformation in X_1–X_2 plane.

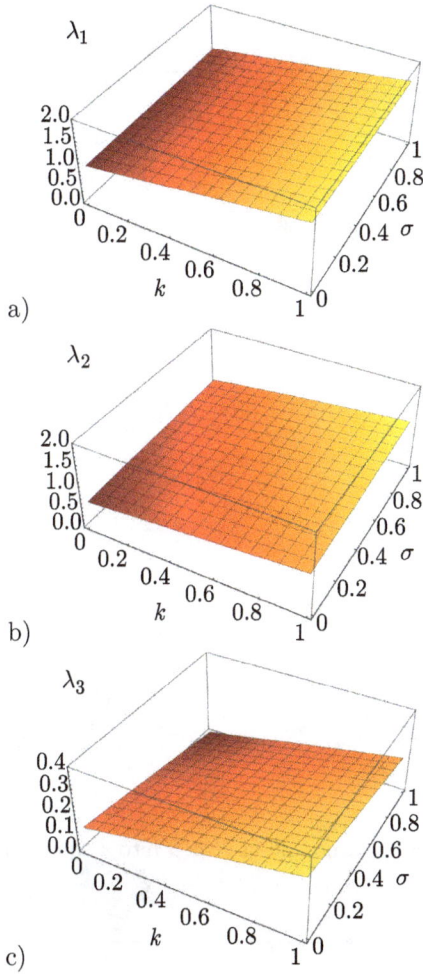

Figure 5.60 Distribution of eigenvalues for octagon 2: a) λ_1; b) λ_2; c) λ_3.

5.4.5 OCTAGON 3

The analysed 2D octagonal module (O3) is presented in Figure 5.62a, with its geometry described in Table 5.17.

The module has two infinitesimal mechanisms and one self-stress state (Figure 5.62b)—self-stress is expressed by relative forces in struts and cables with a multiplier S_0.

The elasticity matrix obtained from the continuum model has a form

$$
\mathbf{E}_{O3} = \frac{EA}{ah}
\begin{bmatrix}
0.6725 + 0.5147k & 0.2242 + 0.1716k & 0 \\
0.2242 + 0.1716k & 0.6725 + 0.5147k & 0 \\
0 & 0 & 0.2242 + 0.1716k
\end{bmatrix}.
$$

$$(5.47)$$

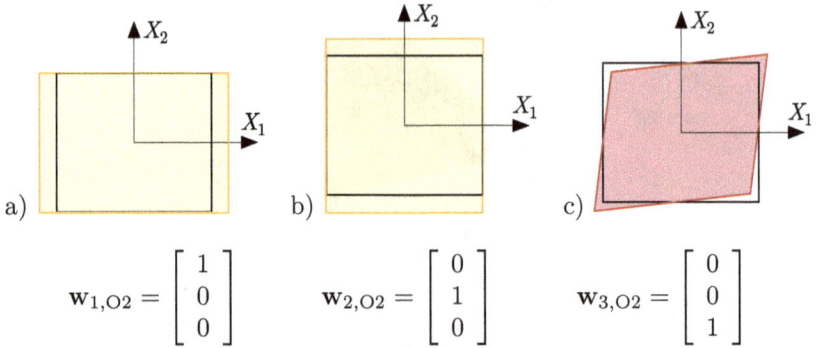

$$\mathbf{w}_{1,O2} = \begin{bmatrix} 1 \\ 0 \\ 0 \end{bmatrix} \qquad \mathbf{w}_{2,O2} = \begin{bmatrix} 0 \\ 1 \\ 0 \end{bmatrix} \qquad \mathbf{w}_{3,O2} = \begin{bmatrix} 0 \\ 0 \\ 1 \end{bmatrix}$$

Figure 5.61 Deformation modes of octagon 2: a) stiff mode corresponding to λ_1; b) soft mode corresponding to λ_2; c) soft mode corresponding to λ_3.

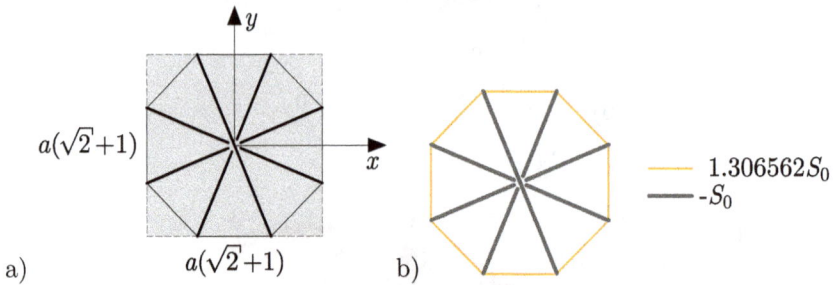

Figure 5.62 Octagon 3: a) module inscribed into a square; b) self-stress state.

Table 5.17
Geometry of octagon 3.

Node No.	x	y	Scheme
1	$-a(1+\sqrt{2})/2$	$-a/2$	
2	$-a(1+\sqrt{2})/2$	$a/2$	
3	$-a/2$	$a(1+\sqrt{2})/2$	
4	$a/2$	$a(1+\sqrt{2})/2$	
5	$a(1+\sqrt{2})/2$	$a/2$	
6	$a(1+\sqrt{2})/2$	$-a/2$	
7	$a/2$	$-a(1+\sqrt{2})/2$	
8	$-a/2$	$-a(1+\sqrt{2})/2$	

It can be noticed that contrary to octagons 1 and 2, matrix $\mathbf{E}$ of octagon 3 does not depend on self-stress forces. Spectral analysis of matrix $\mathbf{E}$ allows us to find the following eigenvalues:

$$\lambda_1 = \frac{0.1177EA}{ah}\left(5.7114 + 4.3713k + \sqrt{3.6245 + 5.5481k + 2.1232k^2}\right) > 0,$$

$$\lambda_2 = \frac{0.1177EA}{ah}\left(5.7114 + 4.3713k - \sqrt{3.6245 + 5.5481k + 2.1232k^2}\right) > 0,$$

$$\lambda_3 = \frac{EA}{ah}(0.2242 + 0.1716k) > 0.$$

$$(5.48)$$

All three eigenvalues are positive and the module has no soft modes of deformation—it should be classified as a nullmode material.

Sections 5.4.1–5.4.5 were focussed on a systematic study on extremal mechanical properties of a series of 2D tensegrity modules. All presented analyses were based on the continuum approach presented in Section 4.2, which can be used for the identification of extremal behaviour of metamaterials. Although 2D systems are considered less practical from the engineering point of view, they help us understand the concept of extremal materials. Identification of extremal mechanical properties of 2D systems leads to mathematically elegant closed-form solutions for eigenvalues that describe these properties. Such solutions are not achievable in 3D systems.

5.5 TENSEGRITY LATTICES IN 2D SPACE

Tensegrity modules described in Section 5.4 can be used to create 2D tensegrity lattices. Such systems are constructed similarly to 3D lattices, which were described in Section 5.3. There are three ways of constructing such systems:

- modular—the lattice is created by joining 2D regular tensegrity modules in nodes,
- non-modular—the lattice might be constructed from 2D regular modules, but a completely new tensegrity system is obtained,
- modular with extra cables—the lattice is created from 2D regular modules joined together using additional cables, which stabilise the whole system.

In Figure 5.63 examples of 2D modular tensegrity lattices are presented: one is based on hexagon 1, another one—on octagon 1. Similarly, other types of tensegrity lattices can be created.

Due to the fact that the basic repetitive cell in such systems is a unit cell, that is a basic tensegrity module, the extremal mechanical properties of

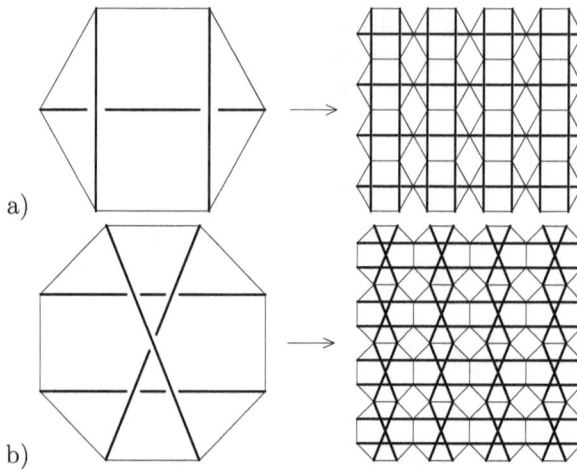

Figure 5.63 2D tensegrity lattices: a) lattice based on hexagon 1; b) lattice based on octagon 1.

the whole metamaterial or metastructure are the same as the properties of single modules described in Section 5.4. The lattice based on hexagon 1 is a unimode system with one soft mode of deformation presented in Figure 5.54c (shear deformation) and two stiff deformation modes depicted in Figures 5.54a and 5.54b (volumetric deformations). The lattice based on octagon 1 is also unimode, with one soft mode of deformation presented in Figure 5.58c (shear deformation) and two stiff deformation modes depicted in Figures 5.58a and 5.58b (volumetric deformations).

In a similar way, other modular 2D tensegrity lattices may be investigated. It is a big advantage of the presented method of analysis, as extremal mechanical properties of complex systems may be identified by analysing very simple structures, such as basic modules. However, it should be noted that such a simplification can be applied only for fully modular systems. Metamaterials or metastructures with non-modular architecture need to be analysed as whole, as they might exhibit a different extremal behaviour than their unit cells.

5.6 FROM METAMATERIAL TO STRUCTURAL SCALE

All examples of 3D and 2D structures presented above in Chapter 5 were based on the continuum description of tensegrity systems. Such a model is accurate for small scales, where the dimensions of single cells tend to zero (in practice it means that they are much smaller than the whole cellular structure). In this study, such a scale is referred to as a material scale. As it is proved above, tensegrity-based metamaterials exhibit many interesting features, such as for example extremal mechanical properties. However, based solely on the continuum analysis of the systems, it cannot be assumed that such properties would also occur in bigger structural scales.

In order to investigate extremal mechanical properties of cellular tensegrity-based lattices in a bigger scale, here referred to as a structural scale, some other approaches should be used. Below, a procedure for determining extremal mechanical properties of tensegrity systems in an arbitrary scale is proposed. In general, the idea consists in the application of the results obtained for the material scale as a starting point for the identification of extremal parameters in the structural scale.

1. Discrete model—arbitrary scale

The structure is analysed using a discrete model (see Section 4.1 for details) and two stiffness matrices are determined: a linear stiffness matrix $\mathbf{K}_L$ and a geometric stiffness matrix $\mathbf{K}_G$. Both matrices depend on two parameters: k and σ, which were described in detail in Section 5.1. Due to very big dimensions of stiffness matrices, it would be very difficult to find the values of parameters k_{extr} and σ_{extr}, for which the structure exhibits extremal mechanical properties.

2. Continuum model—material scale

The structure is analysed using a continuum model (see Section 4.2 for details) and the elasticity matrix $\mathbf{E}$ is determined. The matrix has dimensions of 6×6 (3D systems) or 3×3 (2D systems), and it depends on two parameters: k and σ. After solving the eigenvalue problem

$$(\mathbf{E} - \lambda \mathbf{I})\,\varepsilon = \mathbf{0}, \qquad (5.49)$$

it is possible to determine the values of parameters k_{extr} and σ_{extr}, which ensure occurrence of extremal mechanical properties, if they exist. If no extremal behaviour occurs in the material scale, there is no need to carry on the analysis, as then, the metastructure would also have no extremal properties. Apart from extremal parameters, various deformation modes may be found using this approach, including soft and stiff modes of deformation. Such a model is correct in the material scale, where $a \to 0$ (a is the dimension of a single cell).

3. Discrete model—arbitrary scale—extremal properties

Parameters k_{extr} and σ_{extr} determined in the continuum analysis are substituted into the stiffness matrices, which were calculated in the first step of this procedure:

$$\begin{aligned}
\mathbf{K}_{L,extr} &= \mathbf{K}_L\left(k_{extr}, \sigma_{extr}\right), \\
\mathbf{K}_{G,extr} &= \mathbf{K}_G\left(k_{extr}, \sigma_{extr}\right).
\end{aligned} \qquad (5.50)$$

Afterwards, a strain energy in the discrete model is analysed, through spectral analysis

$$(\mathbf{K}_{\mathrm{L,extr}} + \mathbf{K}_{\mathrm{G,extr}} - \chi \mathbf{I}) \mathbf{q} = \mathbf{0}. \tag{5.51}$$

Extremal properties of the structure can be found by analysing the values of the strain energy: all eigenvalues which are zero (apart from the zero values corresponding to rigid movements) indicate soft modes of deformation. Such deformation modes are expressed by eigenvectors $\mathbf{q}_{\mathrm{soft}}$, which generate the strain energy equal zero. The following equation is satisfied:

$$\frac{1}{2} \mathbf{q}_{\mathrm{soft}}^{\mathrm{T}} (\mathbf{K}_{\mathrm{L,extr}} + \mathbf{K}_{\mathrm{G,extr}}) \mathbf{q}_{\mathrm{soft}} = 0. \tag{5.52}$$

If for a certain pair of parameters k_{extr} and σ_{extr} the scale effect is too big due to the approximation of the continuum model, the eigenvalue corresponding to the soft deformation mode will not be zero, but significantly smaller than others. In this case, one can look for a new pair of parameters $\tilde{k}_{\mathrm{extr}}$ and $\tilde{\sigma}_{\mathrm{extr}}$, that are close to the previous ones, and lead to the zero value of the strain energy. Searching for pairs of extremal parameters close to the already identified ones is much easier and less time-consuming than searching within the whole (k, σ) space.

The procedure described above can be used to identify extremal mechanical properties of tensegrity systems in various scales. It should be taken into account that such extreme behaviour depends on parameters k_{extr} and σ_{extr}, which are found in theoretical models of the systems. It might happen that the determined parameters are unreal from the practical point of view. In this case, the following approach is proposed.

The analysis should be performed for a real, practical value of parameter k and using this value, parameter $\sigma \in (0, \sigma_{\mathrm{extr}})$ should be searched for, which would lead to energy states that are close to zero, but do not reach zero. In this way, semi-soft states could be identified, which would be realisable from the practical point of view.

Below, the proposed procedure is applied to two examples of tensegrity modules: hexagon 1 (2D) and four-strut simplex (3D), in order to demonstrate how it works in practice and to prove that the identified extremal properties of cellular tensegrity metamaterials also occur in a structural scale.

5.6.1   2D EXAMPLE

The first example is a 2D tensegrity module—hexagon 1 described in Section 5.4.1. The module is shown in Figure 5.64, with numbered degrees of freedom and support conditions.

Geometry of the module was described in Section 5.4.1 in Table 5.13. The module has one infinitesimal mechanism (Figure 5.65) and one self-stress state, which was shown in Section 5.4.1 in Figure 5.52b.

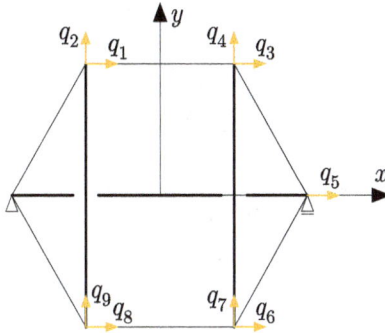

Figure 5.64 Hexagon 1 with numbered degrees of freedom and supports.

In the following, the module is analysed using the procedure described in the previous section.

1. Discrete model within the 2nd order theory—arbitrary scale

The structure is analysed using a discrete model and two stiffness matrices are determined:

$$
\mathbf{K}_{\text{L,H1}} = \frac{EA}{4ah}
\begin{bmatrix}
5k & \sqrt{3}k & -4k & 0 & 0 & 0 & 0 & 0 & 0 \\
\sqrt{3}k & 3k+\frac{4}{\sqrt{3}} & 0 & 0 & 0 & 0 & 0 & 0 & -\frac{4}{\sqrt{3}} \\
-4k & 0 & 5k & -\sqrt{3}k & -k & 0 & 0 & 0 & 0 \\
0 & 0 & -\sqrt{3}k & 3k+\frac{4}{\sqrt{3}} & \sqrt{3}k & 0 & -\frac{4}{\sqrt{3}} & 0 & 0 \\
0 & 0 & -k & \sqrt{3}k & 3k+\frac{4}{\sqrt{3}} & -k & -\sqrt{3}k & 0 & 0 \\
0 & 0 & 0 & 0 & -k & 5k & \sqrt{3}k & -4k & 0 \\
0 & 0 & 0 & -\frac{4}{\sqrt{3}} & -\sqrt{3}k & \sqrt{3}k & 3k+\frac{4}{\sqrt{3}} & 0 & 0 \\
0 & 0 & 0 & 0 & 0 & -4k & 0 & 5k & -\sqrt{3}k \\
0 & -\frac{4}{\sqrt{3}} & 0 & 0 & 0 & 0 & 0 & -\sqrt{3}k & 3k+\frac{4}{\sqrt{3}}
\end{bmatrix} ,
$$

$$(5.53)$$

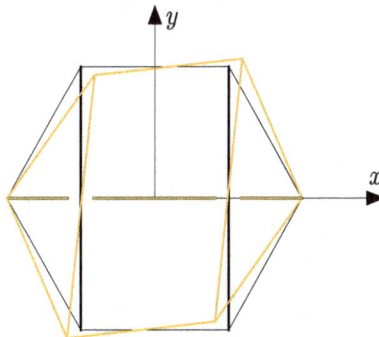

Figure 5.65 Infinitesimal mechanism of hexagon 1.

$$\mathbf{K}_{G,H1} = \frac{S_0}{2ah} \begin{bmatrix} \frac{1}{\sqrt{3}} & -1 & 0 & 0 & 0 & 0 & 0 & \frac{2}{\sqrt{3}} & 0 \\ -1 & \sqrt{3} & 0 & -\frac{2}{\sqrt{3}} & 0 & 0 & 0 & 0 & 0 \\ 0 & 0 & \frac{1}{\sqrt{3}} & 1 & -\sqrt{3} & \frac{2}{\sqrt{3}} & 0 & 0 & 0 \\ 0 & -\frac{2}{\sqrt{3}} & 1 & \sqrt{3} & -1 & 0 & 0 & 0 & 0 \\ 0 & 0 & -\sqrt{3} & -1 & 2\sqrt{3} & -\sqrt{3} & 1 & 0 & 0 \\ 0 & 0 & \frac{2}{\sqrt{3}} & 0 & -\sqrt{3} & \frac{1}{\sqrt{3}} & -1 & 0 & 0 \\ 0 & 0 & 0 & 0 & 1 & -1 & \sqrt{3} & 0 & -\frac{2}{\sqrt{3}} \\ \frac{2}{\sqrt{3}} & 0 & 0 & 0 & 0 & 0 & 0 & \frac{1}{\sqrt{3}} & 1 \\ 0 & 0 & 0 & 0 & 0 & 0 & -\frac{2}{\sqrt{3}} & 1 & \sqrt{3} \end{bmatrix}.$$

$$(5.54)$$

Both matrices depend on two parameters: k and $S_0 = \sigma \cdot EA$.

2. Continuum model—material scale with $a \to 0$

The structure is analysed using the continuum model and the elasticity matrix $\mathbf{E}$ is determined:

$$\mathbf{E}_{H1} = \frac{EA}{ah} \frac{1}{8\sqrt{3}} \begin{bmatrix} 8+9k & 3k & 0 \\ 3k & 8+9k & 0 \\ 0 & 0 & 3k \end{bmatrix} + \frac{S_0}{ah} \frac{1}{4} \begin{bmatrix} 1 & -1 & 0 \\ -1 & 1 & 0 \\ 0 & 0 & -1 \end{bmatrix}.$$

$$(5.55)$$

After solving the eigenvalue problem $(\mathbf{E} - \lambda\mathbf{I})\,\varepsilon = \mathbf{0}$, three eigenvalues depending on parameters k and σ are obtained (the eigenvalues are presented in Section 5.4.1). Soft mode of deformation is identified for the following pairs of parameters k_{extr} and σ_{extr}:

$$\sigma_{\text{extr}} = \frac{k_{\text{extr}}\sqrt{3}}{2}. \qquad (5.56)$$

Hexagon 1 is a unimode material with the following eigenvalues and corresponding eigenvectors that describe deformation modes:

$$\lambda_1 = \frac{EA}{2ah}(\sqrt{3}k + 2): \quad \mathbf{w}_{1,H1} = \begin{bmatrix} 0 \\ 1 \\ 0 \end{bmatrix},$$

$$\lambda_2 = \frac{EA}{2ah\sqrt{3}}(3k + 2): \quad \mathbf{w}_{2,H1} = \begin{bmatrix} 1 \\ 0 \\ 0 \end{bmatrix}, \qquad (5.57)$$

$$\lambda_3 = 0: \quad \mathbf{w}_{3,H1} = \begin{bmatrix} 0 \\ 0 \\ 1 \end{bmatrix}.$$

3. Discrete model within the 2nd order theory—arbitrary scale—extremal properties

Parameters k_{extr} and σ_{extr} determined in the continuum analysis are substituted into stiffness matrices $\mathbf{K}_L$ (Eq. 5.53) and $\mathbf{K}_G$ (Eq. 5.54). Let $k_{extr} = 0.1$, then from Eq. 5.56: $\sigma_{extr} = \sqrt{3}/20$. The following stiffness matrices are obtained:

$$
\mathbf{K}_{L,extr,H1} = \frac{EA}{40ah}
\begin{bmatrix}
5 & \sqrt{3} & -4 & 0 & 0 & 0 & 0 & 0 & 0 \\
\sqrt{3} & 3+\frac{40}{\sqrt{3}} & 0 & 0 & 0 & 0 & 0 & 0 & -\frac{40}{\sqrt{3}} \\
-4 & 0 & 5 & -\sqrt{3} & -1 & 0 & 0 & 0 & 0 \\
0 & 0 & -\sqrt{3} & 3+\frac{40}{\sqrt{3}} & \sqrt{3} & 0 & -\frac{40}{\sqrt{3}} & 0 & 0 \\
0 & 0 & -1 & \sqrt{3} & 3+\frac{40}{\sqrt{3}} & -1 & -\sqrt{3} & 0 & 0 \\
0 & 0 & 0 & 0 & -1 & 5 & \sqrt{3} & -4 & 0 \\
0 & 0 & 0 & -\frac{40}{\sqrt{3}} & -\sqrt{3} & \sqrt{3} & 3+\frac{40}{\sqrt{3}} & 0 & 0 \\
0 & 0 & 0 & 0 & 0 & -4 & 0 & 5 & -\sqrt{3} \\
0 & -\frac{40}{\sqrt{3}} & 0 & 0 & 0 & 0 & 0 & -\sqrt{3} & 3+\frac{40}{\sqrt{3}}
\end{bmatrix},
$$

$$(5.58)$$

$$
\mathbf{K}_{G,extr,H1} = \frac{EA}{20ah}
\begin{bmatrix}
1 & -\sqrt{3} & 0 & 0 & 0 & 0 & 0 & 2 & 0 \\
-\sqrt{3} & 3 & 0 & -2 & 0 & 0 & 0 & 0 & 0 \\
0 & 0 & 1 & \sqrt{3} & -3 & 2 & 0 & 0 & 0 \\
0 & -2 & \sqrt{3} & 3 & -\sqrt{3} & 0 & 0 & 0 & 0 \\
0 & 0 & -3 & -\sqrt{3} & 6 & -3 & \sqrt{3} & 0 & 0 \\
0 & 0 & 2 & 0 & -3 & 1 & -\sqrt{3} & 0 & 0 \\
0 & 0 & 0 & 0 & \sqrt{3} & -\sqrt{3} & 3 & 0 & -2 \\
2 & 0 & 0 & 0 & 0 & 0 & 0 & 1 & \sqrt{3} \\
0 & 0 & 0 & 0 & 0 & 0 & -2 & \sqrt{3} & 3
\end{bmatrix}.
$$

$$(5.59)$$

Spectral analysis $(\mathbf{K}_{L,extr,H1} + \mathbf{K}_{G,extr,H1} - \chi \mathbf{I})\,\mathbf{q}_{H1} = \mathbf{0}$ leads to the following eigenvalues and corresponding eigenvectors:

$$\chi_1 = 1.3547\frac{EA}{ah} :$$
$$\mathbf{q}_{1,H1} = \begin{bmatrix} 0 & 1 & 0 & -1 & 0 & 0 & 1 & 0 & -1 \end{bmatrix}^{\mathrm{T}},$$

$$\chi_2 = 1.2547\frac{EA}{ah} :$$
$$\mathbf{q}_{2,H1} = \begin{bmatrix} 0 & 1 & 0 & 1 & 0 & 0 & -1 & 0 & -1 \end{bmatrix}^{\mathrm{T}},$$

$$(5.60)$$

$$\chi_3 = 0.7385\frac{EA}{ah} :$$
$$\mathbf{q}_{3,H1} = \begin{bmatrix} 0.0357 & 0 & -0.1923 & 0 & 1 & -0.1923 & 0 & 0.0357 & 0 \end{bmatrix}^{\mathrm{T}},$$

$$\chi_4 = 0.2790\frac{EA}{ah} :$$

$$\mathbf{q}_{4,H1} = \begin{bmatrix} 1 & 0 & -0.7903 & 0 & -0.3755 & -0.7903 & 0 & 1 & 0 \end{bmatrix}^{\mathrm{T}},$$

$$\chi_5 = 0.2\frac{EA}{ah}:$$
$$\mathbf{q}_{5,H1} = \begin{bmatrix} 0 & 1 & 0 & -1 & 0 & 0 & -1 & 0 & 1 \end{bmatrix}^{\mathrm{T}},$$

$$\chi_6 = 0.2\frac{EA}{ah}:$$
$$\mathbf{q}_{6,H1} = \begin{bmatrix} 1 & 0 & -1 & 0 & 0 & 1 & 0 & -1 & 0 \end{bmatrix}^{\mathrm{T}},$$

$$\chi_7 = 0.1\frac{EA}{ah}:$$
$$\mathbf{q}_{7,H1} = \begin{bmatrix} 0 & 1 & 0 & 1 & 0 & 0 & 1 & 0 & 1 \end{bmatrix}^{\mathrm{T}},$$

$$\chi_8 = 0.0825\frac{EA}{ah}:$$
$$\mathbf{q}_{8,H1} = \begin{bmatrix} 0.8511 & 0 & 1 & 0 & 0.3239 & 1 & 0 & 0.8511 & 0 \end{bmatrix}^{\mathrm{T}},$$

$$\chi_9 = 0:$$
$$\mathbf{q}_{9,H1} = \begin{bmatrix} 1 & 0 & 1 & 0 & 0 & -1 & 0 & -1 & 0 \end{bmatrix}^{\mathrm{T}}.$$

It can be noticed that one eigenvalue χ_9 equals zero, which proves that the structure exhibits extremal mechanical properties. The corresponding eigenvector describes the soft mode of deformation $\mathbf{q}_9 = \mathbf{q}_{\text{soft}}$, which is presented in Figure 5.66. The strain energy for $\mathbf{q}_9$ equals zero, in accordance with Eq. 5.52.

It is worth noticing that the infinitesimal mode (Figure 5.65) differs from the soft mode of deformation (Figure 5.66). In the soft mode, extensions of the bars are observed, which are leveled by self-stress forces. In the infinitesimal mode, the bars do not extend, but at the same time no self-stress is considered.

In this particular example of hexagon 1, parameters k_{extr} and σ_{extr} determined in the continuum model are consistent with the parameters in an arbitrary scale. It means that both a metamaterial and a metastructure based

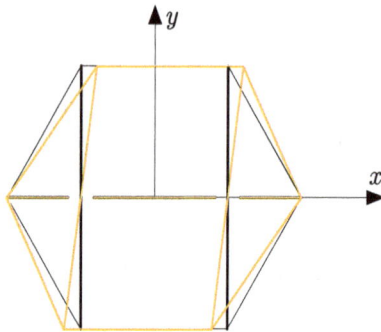

Figure 5.66 Soft deformation mode of hexagon 1.

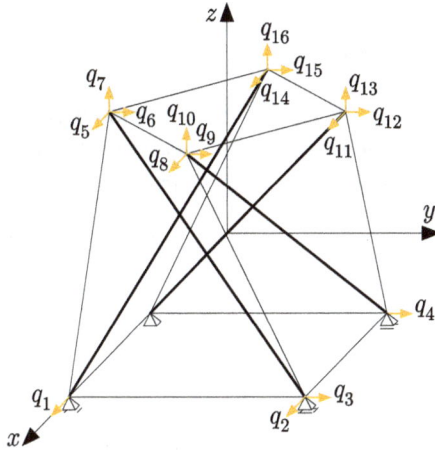

Figure 5.67 Four-strut simplex with numbered degrees of freedom and supports.

on the 2D hexagonal module exhibit extremal mechanical properties for the same pairs of parameters k and σ.

5.6.2 3D EXAMPLE

The second example is a 3D tensegrity module—four-strut simplex described in Section 5.2.2. The module is shown in Figure 5.67, with numbered degrees of freedom and support conditions.

Geometry of the module was described in Section 5.2.2 in Table 5.3. The module has one infinitesimal mechanism (Figure 5.68) and one self-stress state, which was shown in Section 5.2.2 in Figure 3.2.

Below, the module is analysed using the procedure described in the previous section.

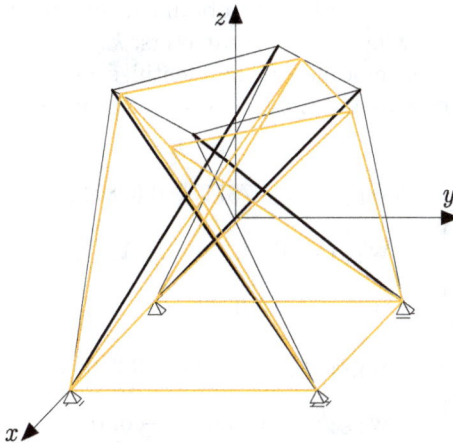

Figure 5.68 Infinitesimal mechanism of the four-strut simplex.

1. Discrete model within the 2nd order theory—arbitrary scale

The structure is analysed using a discrete model and two stiffness matrices are determined: $\mathbf{K}_{L,S4}$ and $\mathbf{K}_{G,S4}$. The matrices cannot be presented here due to their big dimensions. Both matrices depend on two parameters: k and σ.

2. Continuum model—material scale with $a \to 0$

The structure is analysed using the continuum model and the elasticity matrix $\mathbf{E}$ is determined:

$$
\mathbf{E}_{S4} =
\begin{bmatrix}
e_{11} & e_{12} & e_{13} & e_{14} & 0 & 0 \\
 & e_{11} & e_{13} & -e_{14} & 0 & 0 \\
 & & e_{33} & 0 & 0 & 0 \\
 & & & e_{12} & 0 & 0 \\
 & & & & e_{13} & 0 \\
\text{sym.} & & & & & e_{13}
\end{bmatrix},
$$

$$
\begin{aligned}
e_{11} &= \frac{2EA}{a^2}(0.314815 + 1.39827 \cdot k - 0.0794978 \cdot \sigma), \\
e_{12} &= \frac{EA}{a^2}(0.296296 + 0.707107 \cdot k - 0.0134742 \cdot \sigma), \\
e_{13} &= \frac{EA}{a^2}(0.740741 + 0.357771 \cdot k + 0.17247 \cdot \sigma), \\
e_{14} &= \frac{EA}{a^2}(-0.222222 - 0.0808452 \cdot \sigma), \\
e_{33} &= \frac{2EA}{a^2}(0.592593 + 1.43108 \cdot k - 0.17247 \cdot \sigma).
\end{aligned}
\tag{5.61}
$$

After solving the eigenvalue problem $(\mathbf{E} - \lambda \mathbf{I})\,\boldsymbol{\varepsilon} = \mathbf{0}$, six eigenvalues depending on parameters k and σ are obtained. Soft mode of deformation is identified for the following pair of parameters: $k_{\text{extr}} = 0.1$ and $\sigma_{\text{extr}} = 0.546$. The four-strut simplex module can be identified as quasi bimode, since one eigenvalue is close to zero and the second is much smaller than the other four:

$$
\begin{aligned}
\lambda_1 &= 2.4650\frac{EA}{a^2} : & \mathbf{w}_{1,S4} &= [0.678717 \quad 0.678717 \quad 1 \quad 0 \quad 0 \quad 0]^{\mathrm{T}}, \\
\lambda_2 &= 0.8707\frac{EA}{a^2} : & \mathbf{w}_{2,S4} &= [0 \quad 0 \quad 0 \quad 0 \quad 1 \quad 0]^{\mathrm{T}}, \\
\lambda_3 &= 0.8707\frac{EA}{a^2} : & \mathbf{w}_{3,S4} &= [0 \quad 0 \quad 0 \quad 0 \quad 0 \quad 1]^{\mathrm{T}}, \\
\lambda_4 &= 0.7915\frac{EA}{a^2} : & \mathbf{w}_{4,S4} &= [-0.810541 \quad 0.810541 \quad 0 \quad 1 \quad 0 \quad 0]^{\mathrm{T}}, \\
\lambda_5 &= 0.0310\frac{EA}{a^2} : & \mathbf{w}_{5,S4} &= [0.616872 \quad -0.616872 \quad 0 \quad 1 \quad 0 \quad 0]^{\mathrm{T}}, \\
\lambda_6 &= 0 : & \mathbf{w}_{6,S4} &= [-0.736684 \quad -0.736684 \quad 1 \quad 0 \quad 0 \quad 0]^{\mathrm{T}}.
\end{aligned}
\tag{5.62}
$$

3. Discrete model within the 2nd order theory—arbitrary scale—extremal properties

In this case, parameters $k_{\text{extr}} = 0.1$ and $\sigma_{\text{extr}} = 0.546$ determined in the continuum analysis are inaccurate due to the approximation of the continuum model, and in the discrete model, the eigenvalue corresponding to the soft mode of deformation is not zero. However, using the parameter $k_{\text{extr}} = 0.1$, a new parameter $\tilde{\sigma}_{\text{extr}} = 0.5405$ can be found, which leads to the zero value of the strain energy. The stiffness matrices $\mathbf{K}_{\text{L,extr,S4}}$ and $\mathbf{K}_{\text{G,extr,S4}}$ cannot be presented here due to their big dimensions.

Spectral analysis $(\mathbf{K}_{\text{L,extr,S4}} + \mathbf{K}_{\text{G,extr,S4}} - \chi\mathbf{I})\,\mathbf{q}_{\text{S4}} = \mathbf{0}$ leads to the following eigenvalues:

$$\chi = \frac{EA}{a^2}\begin{bmatrix} 1.4133 \\ 1.3075 \\ 1.1804 \\ 1.0297 \\ 0.5698 \\ 0.5579 \\ 0.3993 \\ 0.3250 \\ 0.3089 \\ 0.2542 \\ 0.1956 \\ 0.1271 \\ 0.0529 \\ 0.0405 \\ 0.0230 \\ 0 \end{bmatrix}. \tag{5.63}$$

It can be noticed that one eigenvalue χ_{16} equals zero, which proves that the structure exhibits extremal mechanical properties. The corresponding eigenvector describes the soft mode of deformation $\mathbf{q}_{16} = \mathbf{q}_{\text{soft}}$, which is presented in Figure 5.69. The strain energy for $\mathbf{q}_{16}$ equals zero, in accordance with Eq. 5.52.

Similarly to the 2D example, the infinitesimal mode of the four-strut simplex (Figure 5.68) differs from the soft mode of deformation (Figure 5.69). In the soft mode, extensions of the bars are observed, which are leveled by self-stress forces. In the infinitesimal mode, the bars do not extend, but at the same time no self-stress is considered.

In this example of the four-strut simplex, parameters k_{extr} and σ_{extr} determined in the continuum model are not fully consistent with the parameters in an arbitrary scale. However, both the metamaterial and the metastructure based on the four-strut simplex module exhibit extremal mechanical properties for similar pairs of parameters k and σ.

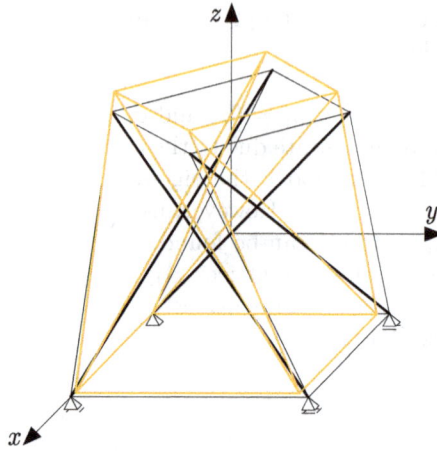

Figure 5.69 Soft deformation mode of the four-strut simplex.

Sections 5.6.1–5.6.2 demonstrated how to apply the procedure for identification of extremal mechanical properties of tensegrity-based metastructures. Two examples of tensegrity modules were presented: hexagon 1 (2D) and four-strut simplex (3D). It was proved that the extremal properties of cellular tensegrity metamaterials also occur in a structural scale.

5.7   TO CONCLUDE

In Chapter 5, a detailed study on extremal mechanical properties of a series of 2D and 3D tensegrity modules and lattices was shown. Presented analyses allow us to draw several conclusions, which are discussed below.

Mechanical behaviour of tensegrity systems can be controlled using two parameters: cable-to-strut stiffness ratio k and self-stress level σ. Both parameters have a significant influence on extremal properties of tensegrity, however, only the self-stress level can be adjusted after the system has been manufactured. The values of k_{extr} and σ_{extr} obtained for most analysed structures are not achievable with the existing materials, as the load-carrying capacity of structural members is not high enough to withstand such a big prestress. However, it was proved in Section 5.2.6 that some of the proposed modules are able to exhibit extremal behaviour using some commonly available typical members.

Extremal behaviour of tensegrity systems can be investigated using either a discrete approach (see Section 4.1) or a continuum model (see Section 4.2). Discrete models may be used for determination of extremal mechanical properties of tensegrity lattices in a structural scale (metastructures). The continuum approach, on the other hand, allows us to study extremal behaviour of tensegrity systems in a material scale (metamaterials).

The elasticity matrix $\mathbf{E}$, which was used in Chapter 5 to evaluate extremal mechanical properties of various tensegrity metamaterials, is an approximation within the first gradient theory [58], and is therefore associated with a scale error. The value of this error depends on the analysed task, including the regularity of the considered structure. It is possible to build a continuum model, which would include higher gradients of displacements, however, the first gradient theory applied in this study is sufficient to identify extremal behaviour of tensegrity metamaterials, where the unit cell is small compared to the size of the whole system.

In the discrete approach, due to very big dimensions of stiffness matrices $\mathbf{K}_L$ and $\mathbf{K}_G$ in most 2D and 3D tasks, it is often difficult, if at all possible, to determine pairs of parameters k_{extr} and σ_{extr} which lead to the occurrence of soft modes of deformation. Therefore, another approach was proposed in Section 5.6, which is based on the consecutive use of the continuum and discrete approach. In the continuum model, such problems do not occur, as the dimensions of the elasticity matrix $\mathbf{E}$ remain the same independent from the task complexity (3×3 in 2D tasks, and 6×6 in 3D tasks). It is therefore easy to find pairs k_{extr} and σ_{extr} leading to extremal properties of the metamaterial.

An important finding is that the analysis of extremal behaviour of complex tensegrity-based metamaterials or metastructures can be substituted with the investigation of extremal mechanical properties of repetitive single modules (cells) or sets of modules (supercells). Big tensegrity lattices have identical mechanical properties (represented by the elasticity matrix $\mathbf{E}$) as their basic cells or supercells.

In Section 5.6, it was demonstrated how to use the pairs of parameters k_{extr} and σ_{extr} determined from the spectral analysis of the elasticity matrix $\mathbf{E}$ to investigate extremal behaviour of tensegrity systems in non-material scales. In the 2D task, k_{extr} and σ_{extr} determined from the continuum model turned out to be accurate in any scale, also the structural one. In the 3D task, k_{extr} and σ_{extr} from the continuum analysis could not be directly applied in bigger scales, but they indicate approximate values of extremal parameters, which are very close to the ones determined from the elasticity matrix $\mathbf{E}$. Therefore, the search for accurate values of extremal properties in the structural scale is easier.

All analyses presented in Chapter 5 were based on the 2nd order theory. They can be used as a reference point for fully non-linear analyses, which are usually applied to study tensegrity systems. However, considering the fact that the values of σ_{extr} in all analysed cases were rather big, the differences between the results obtained according to the 2nd order and the fully nonlinear theory are negligible [4]. The biggest differences between these two theories are noticed when self-stress forces are low. With an increase of prestress in the structure, both theories start to converge. This is why the analyses demonstrated in this study were carried out using the simpler 2nd order theory.

6 Technology and Applications

6.1 TECHNOLOGICAL ASPECTS IN VARIOUS SCALES

Tensegrity lattices described in this book can be constructed using various techniques, which mainly depend on the applied scale. Other methods will be used for tensegrity-based mechanical metamaterials, other for macro-scale lattices. In a material scale (metamaterials), most fabrication techniques are based on additive manufacturing methods. In bigger scales, standard engineering solutions may be applied, such as commonly available rods and cables.

Tensegrity lattices in bigger scales than a metamaterial microstructure can be fabricated using standard structural elements, such as rods or tubes with small cross-sections and commonly available cables. Various parent materials may be used, depending on the required physical and mechanical properties of the metastructure, e.g. steel, aluminium, carbon fiber reinforced polymer (CFRP) or other composites. The biggest challenge in constructing such lattices is a proper design and fabrication of joints. In tensegrity-based systems, structural elements connected in nodes come from various directions, sometimes two or even more struts need to be joined with cables in one node. Moreover, the joints should be constructed in such a way that the axes of all joined elements meet in one point, which is often a very difficult task. Tensegrity joints are usually fabricated using additive manufacturing techniques, as they always need to be designed and produced for each system individually.

Additive manufacturing (also known as 3D printing) is a technology that allows for creation of real objects based on a computer aided design (CAD) or a digital 3D model. The history of this technology dates back to the early 1980s, when Hideo Kodama patented the first rapid prototyping device [77]. The first commercial method for rapid prototyping, called Stereolithography (SLA), was invented by Charles Hull in 1984 [69]. In stereolithography the layers are added by curing photopolymers with ultraviolet (UV) lasers. Hull defined the process as a *system for generating three-dimensional objects by creating a cross-sectional pattern of the object to be formed* [69]. Since then, thanks to the technological progress the additive manufacturing techniques have undergone many transformations and nowadays, various methods are commonly available and new techniques are still being developed.

Additive manufacturing techniques used for metamaterials fabrication can be divided into three main categories [130]:

- ink-based techniques,
- light-based techniques,
- powder-based techniques.

Ink-based techniques consist in a direct deposition of a parent material on a flat substrate, for example, by melting a filament and extruding it through a nozzle. Within this category, two techniques are most commonly used: Fused Deposition Modelling (FDM) and Direct Ink Writing (DIW). In FDM [55] a thermoplastic material is heated until it melts and then, it is extruded through a heated nozzle, it cools down and solidifies. The print quality depends strongly on the filament used, especially its mechanical and thermal properties, as rapid cooling can cause shrinkage of the deposited material. The biggest problem that needs to be faced while using this technique is a weak bond between consecutive layers, which strongly affects the mechanical properties of the printed object. The second mentioned technique, namely DIW [84], uses viscoelastic or viscoplastic ink in a syringe that is extruded through a nozzle under external pressure. The main disadvantage of this method is that it produces rather soft structures that need to be post-processed to gain higher stiffness.

Light-based techniques use a photopolymerisation process, which allows for fabrication of finer features than the ink- and powder-based methods. They use either a UV light or a laser source, which are directed into a photopolymer resin. The oldest and, at the same time, the most popular light-based technique is Stereolithography (SLA) [67], which fabricates structures by selective solidification of the liquid resin through a photopolymerisation reaction. Another light-based technique that can be used to fabricate mechanical metamaterials is Two-Photon Lithography (TPL), also known as Direct Laser Writing. TPL [66] uses a nonlinear dependency of the polymerisation rate on the irradiating light intensity to fabricate 3D microstructures with high resolution.

Powder-based techniques usually use a laser source to fuse powder particles together, either by melting or sintering. To most important factors influencing the print quality belong: shape, size, and distribution of the applied powder. One of the best known powder-based methods is Selective Laser Sintering (SLS) [54], which uses a high-power heating source (laser) to sinter or melt a powder polymer, resin or metal in a heated powder bed, layer after layer. Another commonly used technique is Selective Laser Melting (SLM) [127]. Its working principle is similar to that used in SLS and it is one of the most popular additive manufacturing methods used for commercial fabrication of metallic structures. In this method, in order to minimise the risk of oxidation, the printing process is usually conducted in a special chamber filled with gas. SLM enables us to produce high-quality structures with good mechanical properties.

Tensegrity lattices can be fabricated using all of the methods described above [106], however, the proper method selection requires taking into account the structural scale, materials, expected mechanical properties, size of the printed structure, etc. Pajunen et al. [106] proved that it is possible to fabricate a tensegrity-based structure, whose geometry maintains key tensegrity characteristics and thus, exhibits a mechanical response similar to a pin-jointed tensegrity. Nevertheless, the studies on printable tensegrity are limited so far and this area needs to be explored more thoroughly before tensegrity-based metamaterials and lattices can be produced on a wider scale. However, the rapid development of various additive manufacturing techniques, that has been observed in the past few years, will soon make it possible to apply such systems in real engineering applications.

6.2 APPLICATION OF EXTREMAL TENSEGRITY METAMATERIALS AND STRUCTURES IN CIVIL ENGINEERING

Tensegrity lattices with extremal mechanical properties have a wide range of potential applications in civil engineering. Depending on the considered scale, they can be used either as metamaterials with a microstructure based on tensegrity modules or as medium- and macro-scale metastructures. Tensegrities owe this wide application potential to their unique physical and mechanical properties, such as high strength-to-weight ratio, tunability, high failure resistance, structural smartness, extremal modes of deformation and many others.

It should be noted that tensegrity systems with extremal mechanical properties should always be selected carefully and designed specifically for the intended application. Depending on the desired effect, structures with appropriate soft and stiff deformation modes should be used, and the applied pattern should ensure the expected mechanical behaviour.

Tensegrity-based mechanical metamaterials are aimed mainly at various dynamic applications, namely acoustic, vibration and shock energy damping systems, including seismic protection, acoustic bands, resilient mats for noise and vibration suppression, etc.

Bauer et al. [18] proved that tensegrity metamaterials can exhibit unprecedented failure resistance with enhanced deformability and hugely increased energy absorption capability when compared to other state-of-the-art lattices. The authors assigned these extraordinary results to delocalised deformation of tensegrity systems due to the discontinuity of their compression members. Pal et al. [107] investigated dynamic properties of cellular tensegrity-based beams, plates and solids, which were earlier proposed by Rimoli and Pal in [117]. They demonstrated that the linear wave propagation properties can be significantly altered by changing the level of self-stress. Moreover, the analyses revealed very low wave velocities compared to the parent material and the existence of flat bands at low frequencies. Pajunen et al. [105] studied dynamic behaviour of tensegrity-inspired lattice structures fabricated using the additive manufacturing. Although a stiff polymer was used as a parent material, the structures remained compliant and elastic up to high levels of prestress. Additionally, the authors demonstrated that the existing band gap can be tuned by adjusting the global prestrain in the structure.

Tensegrity metastructures with extremal mechanical properties have not been applied in civil engineering so far, as this is a novel concept proposed in this book. There is, however, a wide range of potential applications of such systems. Extremal metastructures would be ideal for all applications, where a high energy absorption of the system is required, e.g. in medium- and macro-scale vibration damping elements, seismic protection systems, etc.

Tensegrity-inspired metastructures could also be used as lightweight structural elements, such as beams, girders or plates, whose inner architecture is based on a tensegrity lattice. Such a solution would make it possible to reduce structural weight, and to obtain enhanced mechanical properties such as for example an adjustable stiffness or a high damping capacity. Another possible application of extremal tensegrity lattices are 3D fillings of structural elements. Hollow girders or columns could be filled with a tensegrity metastructure, with an aim of improving stiffness and energy absorption capability, and controlling mechanical properties of the system through the adjustment of prestressing forces in cables and struts.

An interesting potential application arising from the development of extremal tensegrity metastructures regards old buildings. The proposed lattices could be used in such structures as reinforcing systems, supporting the original structural elements. An undoubted advantage of such a reinforcing system would be the possibility of controlling structural health of the building, and making necessary adjustments in mechanical behaviour of the lattice. Structural health monitoring system could be integrated with the system responsible for structural control of the tensegrity metastructure, and in this way, the whole structure of the building would be monitored.

Furthermore, tensegrity-based lattices could be applied in adaptive and deployable systems, which work under predefined loading conditions. Such systems would combine two major advantages of tensegrity—extremal mechanical behaviour under certain stress conditions and deployment possibilities. In such applications, however, it would have to be ensured that the particular deployment stages do not interfere with the desired extremal properties of the system.

Together with a rapid development of modern technologies, including the ones used in material sciences, a tendency to search for novel, unconventional engineering solutions has been observed in recent years. Therefore, it is safe to assume that the practice may soon catch up with the theory. Such a perspective justifies a theoretical search for high-performance materials and structures, that hopefully, might soon be applied in civil engineering, and perhaps other fields of applied science.

References

1. B. Adam and I. Smith. *Learning, Self-Diagnosis and Multi-Objective Control of an Active Tensegrity Structure*, volume 140, pages 439–448. Springer, 2006.

2. I. Ahmad. Smart structures and materials. In *ARO Smart Materials, Structures & Mathematical Issues Workshop Proceedings*, pages 13–16, 1988.

3. G. Akhras. Smart materials and smart systems for the future. *Canadian Military Journal*, 1:25–32, 2000.

4. A. Al Sabouni-Zawadzka. Active control of smart tensegrity structures. *Archives of Civil Engineering*, 60(4), 2014.

5. A. Al Sabouni-Zawadzka. Extreme mechanical properties of regular tensegrity unit cells in 3D lattice metamaterials. *Materials*, 13(21):4845, 2020.

6. A. Al Sabouni-Zawadzka and W. Gilewski. Technical coefficients in continuum models of an anisotropic tensegrity module. *Procedia Engineering*, 111:871–876, 2015.

7. A. Al Sabouni-Zawadzka and W. Gilewski. Inherent properties of smart tensegrity structures. *Applied Sciences*, 8(5):787, 2018.

8. A. Al Sabouni-Zawadzka and W. Gilewski. Smart metamaterial based on the simplex tensegrity pattern. *Materials*, 11(5):673, 2018.

9. A. Al Sabouni-Zawadzka and W. Gilewski. Soft and stiff simplex tensegrity lattices as extreme smart metamaterials. *Materials*, 12(1):187, 2019.

10. A. Al Sabouni-Zawadzka and A. Zawadzki. Simulation of a deployable tensegrity column based on the finite element modeling and multibody dynamics simulations. *Archives of Civil Engineering*, 66(4):543–560, 2020.

11. A. Amendola, A. Krushynska, C. Daraio, N. Pugno, and F. Fraternali. Tuning frequency band gaps of tensegrity mass-spring chains with local and global prestress. *International Journal of Solids and Structures*, 155:47–56, 2018.

12. E.H. Anderson and J.M. Sater. SPIE Smart Structures Product Implementation Award: a review of the first ten Years. In *Industrial and Commercial Applications of Smart Structures Technologies 2007*, volume 6527, pages 1–13, 2007.

13. S. Arabnejad and D. Pasini. Mechanical properties of lattice materials via asymptotic homogenization and comparison with alternative homogenization methods. *International Journal of Mechanical Sciences*, 77:249–262, 2013.

14. H. Askes and E. Aifantis. Gradient elasticity in statics and dynamics: An overview of formulations, length scale identification procedures, finite element implementations and new results. *International Journal of Solids and Structures*, 48(13):1962–1990, 2011.

15. E. Barchiesi, M. Spagnuolo, and L. Placidi. Mechanical metamaterials: a state of the art. *Mathematics and Mechanics of Solids*, 24(10):108128651773569, 2018.

16. D.L. Barnes, W. Miller, K.E. Evans, and A. Marmier. Modelling negative linear compressibility in tetragonal beam structures. *Mechanics of Materials*, 46:123–128, 2012.

17. K.J. Bathe. *Finite Element Procedures in Engineering Analysis*. Prentice-Hall, 1996.

18. J. Bauer, J. Kraus, C. Crook, J. Rimoli, and L. Valdevit. Tensegrity metamaterials: toward failure-resistant engineering systems through delocalized deformation. *Advanced Materials*, 33(10):2005647, 2021.

19. N. Bel Hadj Ali and I. Smith. Dynamic behavior and vibration control of a tensegrity structure. *International Journal of Solids and Structures*, 47(9), 2010.

20. N. Ben Kahla. Equivalent beam-column analysis of guyed towers. *Computers & Structures*, 55(4):631–645, 1995.

21. K. Bertoldi, P. Reis, S. Willshaw, and T. Mullin. Negative Poisson's ratio behavior induced by an elastic instability. *Advanced Materials*, 22(3):361–366, 2010.

22. K. Bertoldi, V. Vitelli, J. Christensen, and M. Hecke. Flexible mechanical metamaterials. *Nature Reviews Materials*, 2(11):17066, 2017.

23. D. Betz, W. Staszewski, G. Thursby, and B. Culshaw. Structural damage identification using multifunctional Bragg grating sensors: II. Damage detection results and analysis. *Smart Materials and Structures*, 15(5):1313, 2006.

24. S. Brule, E. Javelaud, S. Enoch, and S. Guenneau. Experiments on seismic metamaterials: molding surface waves. *Physical Review Letters*, 112:133901, 2014.

25. L. Cabras and M. Brun. Auxetic two-dimensional lattice with Poisson's Ratio arbitrarily close to −1. In *Proceedings of the Royal Society A: Mathematical, Physical and Engineering Sciences*, volume 470, pages 1–23, 2014.

26. M. Cai, X. Liu, G. Hu, and P. Zhou. Customization of two-dimensional extremal materials. *Materials & Design*, 218:110657, 2022.

27. C.R. Calladine. Buckminster Fuller's "Tensegrity" structures and Clerk Maxwell's rules for the construction of stiff frames. *International Journal of Solids and Structures*, 14(2):161–172, 1978.

28. C.R. Calladine and S. Pellegrino. First-order infinitesimal mechanisms. *International Journal of Solids and Structures*, 27(4):505–515, 1991.

29. F. Carrion, J. Doyle, and A. Lozano. Structural health monitoring and damage detection using a sub-domain inverse method. *Smart Materials and Structures*, 12:776, 2003.

30. G. Cazzulani, S. Cinquemani, and L. Comolli. Enhancing active vibration control performances in a smart structure by using fiber Bragg gratings sensors. In *Sensors and Smart Structures Technologies for Civil, Mechanical, and Aerospace Systems 2012*, volume 8345, pages 849–857, 2012.

31. G. Cazzulani, S. Cinquemani, L. Comolli, and A. Gardella. Reducing vibration in carbon fiber structures with piezoelectric actuators and fiber Bragg grating sensors. In *Proceedings of SPIE – The International Society for Optical Engineering*, volume 8341, 2012.

32. P. Chadwick, M. Vianello, and S. Cowin. A new proof that the number of linear elastic symmetries is eight. *Journal of the Mechanics and Physics of Solids*, 49(11):2471–2492, 2001.

33. C.M. Chang and B. Spencer. An experimental study of active base isolation control for seismic protection. In *Proceedings of SPIE – The International Society for Optical Engineering*, volume 7647, 2010.

34. H. Chen and C. Chan. Acoustic cloaking in three dimensions using acoustic metamaterials. *Applied Physics Letters*, 91:183518, 2007.

35. M.A. Crisfield. *Non-Linear Finite Element Analysis of Solids and Structures*, volume 1: Essentials. Wiley, 2003.

36. T.J. Cui, D.R. Smith, and R. Liu. *Metamaterials: Theory, Design, and Applications*. Springer, 2010.

37. D. De Tommasi, G. Marano, G. Puglisi, and F. Trentadue. Morphological optimization of tensegrity-type metamaterials. *Composites Part B Engineering*, 115:182–187, 2016.

38. J.O. Dow, Z.W. Su, C.C. Feng, and C. Bodley. Equivalent continuum representation of structures composed of repeated elements. *AIAA Journal*, 23(10):1564–1569, 1985.

39. K. Dudek, D. Attard, R. Caruana-Gauci, K. Wojciechowski, and J. Grima. Unimode metamaterials exhibiting negative linear compressibility and negative thermal expansion. *Smart Materials and Structures*, 25:025009, 2016.

40. D.G. Emmerich. Construction de réseaux autotendants. *US Patent 1,377,290*, 1964.

41. N. Engheta and R.W. Ziolkowski. *Metamaterials: Physics and Engineering Explorations*. Wiley-IEEE Press, 2006.

42. A.C. Eringen. *Microcontinuum Field Theories: I. Foundations and Solids.* Springer New York, 1999.

43. K.E. Evans. Auxetic polymers: a new range of materials. *Endeavour,* 15(4):170–174, 1991.

44. F. Fabbrocino, G. Carpentieri, A. Amendola, R. Penna, and F. Fraternali. Accurate numerical methods for studying the nonlinear wave-dynamics of tensegrity metamaterials. In *Eccomas Procedia COMPDYN 2017,* pages 3911–3922, 2017.

45. H. Fang, S. Li, H. Ji, and K.W. Wang. Uncovering the deformation mechanisms of origami metamaterials by introducing generic degree-four vertices. *Physical Review E,* 94(4):043002, 2016.

46. M. Farshad. Intelligent materials and structures. *Scientia Iranica,* 2(1), 1995.

47. E. Filipov, T. Tachi, and G. Paulino. Origami tubes assembled into stiff, yet reconfigurable structures and metamaterials. *Proceedings of the National Academy of Sciences of the United States of America,* 112(40), 2015.

48. B. Florijn, C. Coulais, and M. Hecke. Programmable mechanical metamaterials. *Physical Review Letters,* 113(17), 2014.

49. S. Forest and K. Sab. Stress gradient continuum theory. *Mechanics Research Communications,* 40:16–25, 2011.

50. F. Fraternali, G. Carpentieri, and A. Amendola. On the mechanical modeling of the extreme softening/stiffening response of axially loaded tensegrity prisms. *Journal of the Mechanics and Physics of Solids,* 74:136–157, 2015.

51. F. Fraternali, G. Carpentieri, A. Amendola, R. Skelton, and V. Nesterenko. Multiscale tunability of solitary wave dynamics in tensegrity metamaterials. *Applied Physics Letters,* 105:201903, 2014.

52. R.B. Fuller. Tensile-integrity structures. *US Patent US3063521A,* 1962.

53. R. Galea, K. Dudek, P.S. Farrugia, L. Mangion, J. Grima, and R. Gatt. Reconfigurable magneto-mechanical metamaterials guided by magnetic fields. *Composite Structures,* 280:114921, 2021.

54. X. Gan, G. Fei, J. Wang, Z. Wang, M. Lavorgna, and H. Xia. *Powder quality and electrical conductivity of selective laser sintered polymer composite components,* pages 149–185. Woodhead Publishing, 2020.

55. X. Gao, N. Yu, and J. Li. *Influence of Printing Parameters and Filament Quality on Structure and Properties of Polymer Composite Components Used in the Fields of Automotive,* pages 303–330. Woodhead Publishing, 2020.

56. W. Gilewski and A. Al Sabouni-Zawadzka. On possible applications of smart structures controlled by self-stress. *Archives of Civil and Mechanical Engineering,* 15(2):469–478, 2015.

57. W. Gilewski and A. Al Sabouni-Zawadzka. Equivalent mechanical properties of tensegrity truss structures with self-stress included. *European Journal of Mechanics - A/Solids*, 83(4):103998, 2020.

58. W. Gilewski and A. Al Sabouni-Zawadzka. Towards recognition of scale effects in a solid model of lattices with tensegrity-inspired microstructure. *Solids*, 2(1):50–59, 2021.

59. W. Gilewski and A. Kasprzak. 3D continuum models of tensegrity modules with the effect of self-stress. In *11th World Congress on Computational Mechanics, 5th European Conference on Computational Mechanics, Barcelona*, 2014.

60. A. Gonzalez, A. Luo, and D. Shi. Reconfiguration of multi-stage tensegrity structures using infinitesimal mechanisms. *Latin American Journal of Solids and Structures*, 16, 2019.

61. G. Greaves, A. Greer, R. Lakes, and T. Rouxel. Poisson's ratio and modern materials. *Nature Materials*, 10(12):986, 2011.

62. A.E. Green and W. Zerna. *Theoretical Elasticity*. Oxford University Press, Oxford, UK, 1968.

63. J. Grima, D. Attard, R. Caruana-Gauci, and R. Gatt. Negative linear compressibility of hexagonal honeycombs and related systems. *Scripta Materialia*, 65:565–568, 2011.

64. J. Grima, D. Attard, and R. Gatt. Truss-type systems exhibiting negative compressibility. *physica status solidi (b)*, 245(11):2405–2414, 2008.

65. Halfen. *Technical Datasheet DT 13-PL.*

66. V. Harinarayana and Y.C. Shin. Two-photon lithography for three-dimensional fabrication in micro/nanoscale regime: a comprehensive review. *Optics & Laser Technology*, 142(2):107180, 2021.

67. J. Huang. A Review of Stereolithography: Processes and Systems. *Processes*, 8(9):1138, 2020.

68. T. Hughes. *The Finite Element Method: Linear Static and Dynamic Finite Element Analysis*, volume 78. Dover Publications, 2000.

69. C. Hull. Apparatus for production of three-dimensional objects by stereolithography. *US Patent 4,575,330*, 1984.

70. M. Kadic, T. Bückmann, R. Schittny, P. Gumbsch, and M. Wegener. Pentamode metamaterials with independently tailored bulk modulus and mass density. *Physical Review Applied*, 2(5):054007, 2014.

71. M. Kadic, T. Bückmann, N. Stenger, M. Thiel, and M. Wegener. On the practicability of pentamode mechanical metamaterials. *Applied Physics Letters*, 100(19):191901, 2012.

72. A. Kasprzak. *Ocena możliwości wykorzystania konstrukcji tensegrity w bu-downictwie mostowym.* PhD thesis, Warsaw University of Technology, Poland, 2014.

73. H. Kazemi and J. Norato. Topology optimization of programmable lattices with geometric primitives. *Structural and Multidisciplinary Optimization,* 65(1), 2022.

74. K. Kebiche, K. Aoual, and R. Motro. Continuum models for systems in a selfstress state. *International Journal of Space Structures,* 23(2):103–115, 2008.

75. R. Khajehtourian and D. Kochmann. Soft adaptive mechanical metamaterials. *Frontiers in Robotics and AI,* 8:673478, 2021.

76. T. Klatt and M. Haberman. A nonlinear negative stiffness metamaterial unit cell and small-on-large multiscale material model. *Journal of Applied Physics,* 114(3), 2013.

77. H. Kodama. Stereoscopic figure drawing device. *Japanese Patent JP S56-144478,* 1981.

78. K. Koohestani and S.D. Guest. A new approach to the analytical and numerical form-finding of tensegrity structures. *International Journal of Solids and Structures,* 50(19):2995–3007, 2013.

79. R. Lakes. Extreme damping in composite materials with a negative stiffness phase. *Physical Review Letters,* 86:2897–900, 2001.

80. A. Lazarus and P.M. Reis. Soft actuation of structured cylinders through auxetic behavior. *Advanced Engineering Materials,* 17(6):815–820, 2015.

81. J.B. Lee, S. Peng, Y.H. Roh, H. Funabashi, N. Park, E. Rice, L. Chen, R. Long, M. Wu, and D. Luo. A mechanical metamaterial made from a DNA hydrogel. *Nature Nanotechnology,* 7(12), 2012.

82. J.H. Lee, J. Singer, and E. Thomas. Micro-/nanostructured mechanical meta-materials. *Advanced Materials,* 24(36):4782–4810, 2012.

83. T.U. Lee, Y. Chen, M.T. Heitzmann, and J.M. Gattas. Compliant curved-crease origami-inspired metamaterials with a programmable force-displacement response. *Materials & Design,* 207(40):109859, 2021.

84. J. Lewis. Direct ink writing of 3D functional materials. *Advanced Functional Materials,* 16(17):2193–2204, 2006.

85. T. Lewiński. On algebraic equations of elastic trusses, frames and grillages. *Journal of Theoretical and Applied Mechanics,* 39:307–322, 2001.

86. J.L. Liu, S. Zhu, Y.L. Xu, and Y. Zhang. Displacement-based design approach for highway bridges with SMA isolators. *Smart Structures and Systems,* 8, 2011.

87. K. Liu, T. Zegard, P.P. Pratapa, and G.H. Paulino. Unraveling tensegrity tessellations for metamaterials with tunable stiffness and bandgaps. *Journal of the Mechanics and Physics of Solids*, 131:147–166, 2019.

88. K.C. Lu, J.H. Weng, and C.H. Loh. Turning the building into a smart structure: integrating health monitoring. In *Proceedings of SPIE – Sensors and Smart Structures Technologies for Civil, Mechanical, and Aerospace Systems*, volume 7292, 2009.

89. Y. Luo, X. Xu, T. Lele, S. Kumar, and D.E. Ingber. A multi-modular tensegrity model of an actin stress fiber. *Journal of Biomechanics*, 41(11):2379–2387, 2008.

90. Y. Ma, Q. Zhang, Y. Dobah, F. Scarpa, F. Fraternali, R. Skelton, D. Zhang, and J. Hong. Meta-tensegrity: Design of a tensegrity prism with metal rubber. *Composite Structures*, 206:644–657, 2018.

91. W. Miller, K. Evans, and A. Marmier. Negative linear compressibility in common materials. *Applied Physics Letters*, 106(23):231903, 2015.

92. G. Milton. Adaptable nonlinear bimode metamaterials using rigid bars, pivots, and actuators. *Journal of the Mechanics and Physics of Solids*, 61(7):1561–1568, 2013.

93. G. Milton. Complete characterization of the macroscopic deformations of periodic unimode metamaterials of rigid bars and pivots. *Journal of the Mechanics and Physics of Solids*, 61(7):1543–1560, 2013.

94. G. Milton and A. Cherkaev. Which elasticity tensors are realizable? *Journal of Engineering Materials and Technology*, 117, 1995.

95. R.D. Mindlin. Micro-structure in linear elasticity. *Archive for Rational Mechanics and Analysis*, 16:51–78, 1964.

96. R.D. Mindlin. Second gradient of strain and surface-tension in linear elasticity. *International Journal of Solids and Structures*, 1(4):417–438, 1965.

97. M. Modano, I. Mascolo, F. Fraternali, and Z. Bieniek. Numerical and Analytical approaches to the self-equilibrium problem of class $\theta = 1$ tensegrity metamaterials. *Frontiers in Materials*, 5(5), 2018.

98. R. Motro. *Tensegrity: Structural Systems for the Future*. Kogan Page Science, 2003.

99. M. Nehdi, M.S. Alam, and M. Youssef. Seismic behaviour of repaired superelastic shape memory alloy reinforced concrete beam-column joint. *Smart Structures and Systems*, 7(5):329–348, 2011.

100. Z. Nicolaou and A. Motter. Mechanical metamaterials with negative compressibility transitions. *Nature Materials*, 11(7):608–13, 2012.

101. A.K. Noor and C.M. Andersen. Analysis of beam-like lattice trusses. *Computer Methods in Applied Mechanics and Engineering*, 20(1):53–70, 1979.

102. A.K. Noor, M.S. Anderson, and W.H. Greene. Continuum Models for beam- and Platelike Lattice Structures. *AIAA Journal*, 16(12):1219–1228, 1978.

103. G. Odegard, T. Gates, L. Nicholson, and K. Wise. Equivalent-continuum modeling of nano-structured materials. *Composites Science and Technology*, 62(14):1869–1880, 2002.

104. J. Overvelde, T. de Jong, Y. Shevchenko, S. Becerra, G. Whitesides, J. Weaver, C. Hoberman, and K. Bertoldi. A three-dimensional actuated origami-inspired transformable metamaterial with multiple degrees of freedom. *Nature Communications*, 7(1), 2016.

105. K. Pajunen, P. Celli, and C. Daraio. Prestrain-induced bandgap tuning in 3D-printed tensegrity-inspired lattice structures. *Extreme Mechanics Letters*, 44:101236, 2021.

106. K. Pajunen, P. Johanns, R.K. Pal, J. Rimoli, and C. Daraio. Design and impact response of 3D-printable tensegrity-inspired structures. *Materials & Design*, 182:107966, 2019.

107. R.K. Pal, M. Ruzzene, and J. Rimoli. Tunable wave propagation by varying prestrain in tensegrity-based periodic media. *Extreme Mechanics Letters*, 22:149–156, 2018.

108. J. Paulose, A. Meeussen, and V. Vitelli. Selective buckling via states of self-stress in topological metamaterials. *Proceedings of the National Academy of Sciences of the United States of America*, 112(25), 2015.

109. S. Pellegrino. Structural computations with the singular value decomposition of the equilibrium matrix. *International Journal of Solids and Structures*, 30(21):3025–3035, 1993.

110. J. Pełczyński and W. Gilewski. An extension of algebraic equations of elastic trusses with self-equilibrated system of forces. In *6th European Conference on Computational Mechanics*, 2018.

111. J. Pełczyński and W. Gilewski. Algebraic formulation for moderately thick elastic frames, beams, trusses, and grillages within Timoshenko Theory. *Mathematical Problems in Engineering*, 2019:9, 2019.

112. D. Polyzos and D.I. Fotiadis. Derivation of Mindlin's first and second strain gradient elastic theory via simple lattice and continuum models. *International Journal of Solids and Structures*, 49(3):470–480, 2012.

113. E. Ptochos and G. Labeas. Elastic modulus and Poisson's ratio determination of micro-lattice cellular structures by analytical, numerical and homogenisation methods. *Journal of Sandwich Structures and Materials*, 14(5):597–626, 2012.

114. J. Qu, K. Muamer, and M. Wegener. Three-dimensional poroelastic metamaterials with extremely negative or positive effective static volume compressibility. *Extreme Mechanics Letters*, 22, 2018.

115. J. Quirant, M.N. Kazi-Aoual, and R. Motro. Designing tensegrity systems: the case of a double layer grid. *Engineering Structures*, 25:1121–1130, 2003.

116. L. Rhode-Barbarigos. *An Active Deployable Tensegrity Structure*. PhD thesis, École Polytechnique Fédérale de Lausanne, Switzerland, 2012.

117. J. Rimoli and R.K. Pal. Mechanical response of 3-dimensional tensegrity lattices. *Composites Part B Engineering*, 115:30–42, 2017.

118. H. Salahshoor, R.K. Pal, and J. Rimoli. Material symmetry phase transitions in three-dimensional tensegrity metamaterials. *Journal of the Mechanics and Physics of Solids*, 119(2):382–399, 2018.

119. A. Salehian and D. Inman. Dynamic analysis of a lattice structure by homogenization: Experimental validation. *Journal of Sound and Vibration*, 316(1-5):180–197, 2008.

120. M. Schenk and S. Guest. Geometry of Miura-folded metamaterials. *Proceedings of the National Academy of Sciences of the United States of America*, 110:3276–3281, 2013.

121. D. Schurig, J.J. Mock, B.J. Justice, S.A. Cummer, J.B. Pendry, A.F. Starr, and D.R. Smith. Metamaterial electromagnetic cloak at microwave frequencies. *Science*, 314(5801):977–980, 2006.

122. J. Shim, S. Shan, A. Košmrlj, S. Kang, E. Chen, J. Weaver, and K. Bertoldi. Harnessing instabilities for design of soft reconfigurable auxetic/chiral materials. *Soft Matter*, 9(34), 2013.

123. A. Silva, F. Monticone, G. Castaldi, V. Galdi, A. Alù, and N. Engheta. Performing mathematical operations with metamaterials. *Science*, 343(6167):160–163, 2014.

124. G. Singh, R. Ni, and A. Marwaha. A review of metamaterials and its applications. *International Journal of Engineering Trends and Technology*, 19(6):305–310, 2015.

125. R. Skelton and M. Oliveira. *Tensegrity Systems*. Springer, 2009.

126. K.D. Snelson. Continuous tension, discontinuous compression structures. *US Patent US3169611A*, 1965.

127. X. Song, W. Zhai, R. Huang, J. Fu, M.W. Fu, and F. Li. *Metal-Based 3D-Printed Micro Parts & Structures*, pages 448–461. Elsevier, 2022.

128. C.M. Soukoulis, S. Linden, and M. Wegener. Negative refractive index at optical wavelengths. *Science*, 315(5808):47–49, 2007.

129. A. Suleman, E. Prasad, R. Blackow, and D. Waechter. *Smart Structures – an Overview*, pages 3–16. Springer Vienna, 2001.

130. J.U. Surjadi, L. Gao, H. Du, X. Li, X. Xiong, N.X. Fang, and Y. Lu. Mechanical metamaterials and their engineering applications. *Advanced Engineering Materials*, 21(3):1800864, 2019.

131. R. Tao, L. Ji, Z. Wan, T. Li, W. Wu, L. Binbin, L. Ma, and D. Fang. 4D printed origami metamaterials with tunable compression twist behavior and stress-strain curves. *Composites Part B Engineering*, 201:108344, 2020.

132. A. Teughels and G. De Roeck. Continuum models for beam- and plate-like lattice structures. In *IASS-IACM 2000, Fourth International Colloquium on Computation of Shells and Spatial Structures*, 2000.

133. G. Tibert. *Deployable Tensegrity Structures for Space Applications*. PhD thesis, Royal Institute of Technology, Sweden, 2002.

134. Titan. *Technical Datasheet AWS 151 Rev.4*.

135. V. Wadhawan. Smart structures and materials. *Resonance*, 10(11):27–41, 2005.

136. S. Waitukaitis, R. Menaut, B. Chen, and M. Hecke. Origami multistabilty: from single vertices to metasheets. *Physical Review Letters*, 114(5):055503, 2015.

137. H.X. Wang and S. Chung. Equivalent elastic constants of truss core sandwich plates. *Journal of Pressure Vessel Technology*, 133(4):041203, 2011.

138. Y. Wang, X. Liu, R. Zhu, and G. Hu. Wave propagation in tunable lightweight tensegrity metastructure. *Scientific Reports*, 8(1):11482, 2018.

139. Y. Wang and C. Wang. Buckling of ultrastretchable kirigami metastructures for mechanical programmability and energy harvesting. *International Journal of Solids and Structures*, 213:93–102, 2021.

140. Y. Wei, X. Liu, and G. Hu. Quadramode materials: Their design method and wave property. *Materials & Design*, 210:110031, 2021.

141. D. Williamson and R. Skelton. A general class of tensegrity structures: topology and prestress equilibrium analysis. *Journal of Guidance, Control, and Dynamics*, 26(5):685–694, 2003.

142. C. Wu, B. Neuner III, J. John, A. Milder, B. Zollars, S. Savoy, and G. Shvets. Metamaterial-based integrated plasmonic absorber/emitter for solar thermo-photovoltaic systems. *Journal of Optics*, 14(2):024005, 2012.

143. L. Wu, Y. Wang, K. Chuang, F. Wu, Q. Wang, W. Lin, and H. Jiang. A brief review of dynamic mechanical metamaterials for mechanical energy manipulation. *Materials Today*, 44(5), 2020.

144. L. Yang, O. Harrysson, H. West, and D. Cormier. Mechanical properties of 3D re-entrant honeycomb auxetic structures realized via additive manufacturing. *International Journal of Solids and Structures*, 69-70:475–490, 2015.

145. A.A. Zadpoor. Mechanical meta-materials. *Materials Horizons*, 3(5):371–381, 2016.

146. A. Zawadzki and A. Al Sabouni-Zawadzka. In search of lightweight deployable tensegrity columns. *Applied Sciences*, 10(23):8676, 2020.

147. Z. Zhai, L. Wu, and H. Jiang. Mechanical metamaterials based on origami and kirigami. *Applied Physics Reviews*, 8(4):41319, 2021.

148. L.Y. Zhang, S.X. Li, S.X. Zhu, B.Y. Zhang, and G. Xu. Automatically assembled large-scale tensegrities by truncated regular polyhedral and prismatic elementary cells. *Composite Structures*, 184:30–40, 2018.

149. Q. Zhang, D. Zhang, Y. Dobah, F. Scarpa, F. Fraternali, and R. Skelton. Tensegrity cell mechanical metamaterial with metal rubber. *Applied Physics Letters*, 113(3):031906, 2018.

150. A. Zhao, Z. Zhao, X. Zhang, X. Cai, L. Wang, T. Wu, and H. Chen. Design and experimental verification of a water-like pentamode material. *Applied Physics Letters*, 110:011907, 2017.

151. X. Zheng, H. Lee, T.H. Weisgraber, M. Shusteff, J. DeOtte, E.B. Duoss, J.D. Kuntz, M.M. Biener, Q. Ge, J.A. Jackson, S.O. Kucheyev, N.X. Fang, and C.M. Spadaccini. Ultralight, ultrastiff mechanical metamaterials. *Science*, 344(6190):1373–1377, 2014.

152. O.C Zienkiewicz, R. Taylor, and J. Zhu. *The Finite Element Method: Its Basis and Fundamentals*, volume I. Butterworth and Heinemann, 2005.

Index

Note: Locators in *italics* represent figures and **bold** indicate tables in the text.

For Product Safety Concerns and Information please contact our EU
representative GPSR@taylorandfrancis.com
Taylor & Francis Verlag GmbH, Kaufingerstraße 24, 80331 München, Germany

www.ingramcontent.com/pod-product-compliance
Lightning Source LLC
Chambersburg PA
CBHW070733220326
41598CB00024BA/3408